Temperate Garden Plant Families

Temperate Garden Plant Families

The Essential Guide *to* Identification *and* Classification

Peter Goldblatt and John C. Manning

Timber Press
Portland, Oregon

Published in 2019 by Timber Press, Inc.
The Haseltine Building
133 S.W. Second Avenue, Suite 450
Portland, Oregon 97204-3527
timberpress.com

Printed in China

Text Design by Sean McCormick

ISBN 978-1-60469-498-7

Catalog records for this book are available from the Library of Congress and the British Library.

CONTENTS

PREFACE

This book is primarily addressed to gardeners, horticulturists, and plant enthusiasts working with plants of temperate climates, thus plants that are hardy or half-hardy. It is not encyclopedic—many families are not be dealt with because they are entirely tropical in distribution or mostly not in cultivation in temperate gardens. It is intended as an introduction to the concept of plant families, especially in light of the more recent changes in the circumscription of many families and the nomenclature of plants at the level of family, genus, and species.

Sources of primary information include first and foremost *Mabberley's Plant-Book, Fourth Edition* (Mabberley 2017), which treats all families and genera according to the most current classification. Wikipedia provided useful ancillary material. Families are recognized and circumscribed largely following the recommendations of the Angiosperm Phylogeny Group (APG) in its latest publications (APG 2003, 2009), the former offering some alternative treatments. We include several modifications to the APG classification as suggested by later authors. *Taxonomy of Vascular Plants* (Lawrence 1951), though arranged according to an out-of-date classification, is nevertheless a source of invaluable details about many plant families. Specialized terms are defined in the Glossary. Additional sources of useful information are included in the References.

Many colleagues helped us refine our entries for families not well known to us: Ihsan Al-Shehbaz (Brassicaceae), Peter Bruyns (Apocynaceae), Peter Hoch (Onagraceae), Cornelia Klak (Aizoaceae), Peter Linder (Poaceae), P. P. Lowry II (Araliaceae), Anthony Magee (Apiaceae), Deirdre Snijman (Amaryllidaceae), and Charlotte Taylor (Rubiaceae). We thank Carmen Ulloa Ulloa for her photographs of *Viola pygmaea*.

INTRODUCTION

What are plant families you ask? The word *family* means different things to different people. To biologists, families are a formal and specific rank of classification. Families include a number of *genera* that share characteristics and are known or believed to be related to one another more closely than to genera of other families.

For the nonscientist it seems legitimate to ask, Why develop a knowledge of plant families at all? In a broad sense it provides a framework for thinking about plants and for arranging hundreds or even thousands of plant names in a coherent and predictive pattern. Knowledge of plant families also allows the reader to develop a deeper understanding of the plants you grow and love. No, knowledge of plant families will not make you a better gardener, but it will provide a deeper appreciation of the plant world and the plants you cultivate. Do you love primulas and primroses, members of the genus *Primula*? Well then, you will surely be fascinated to know that shooting stars, the genus *Dodecatheon*, belong to the same family, Primulaceae. Cyclamens, the genus *Cyclamen*, are also members of the Primulaceae. This may give you pause, Why? Look at the arrangement of the flowers—both primulas and shooting stars have flowers clustered together at a single point on the stem, an arrangement called an umbel. Also, the petals are joined together at the base, and the fruits are more or less identical to those of cyclamens. The fruits are dry and of the type known technically as capsules, and in these genera they split open in a circle near the top of the ripe fruit, a feature rare or absent in other plant families. The genus *Lysimachia*, which includes gooseneck loosestrife, yellow loosestrife, and creeping jenny among others, has similar capsules, but often the flowers are arranged in different ways on the stem. And as common names go, they are often confusing. Thus purple loosestrife is a member of a different family, Lythraceae (the loosestrife or crepe myrtle family); common names are frequently a poor guide to family relationships.

Many other Primulaceae are trees, most of them tropical, and so temperate gardeners have probably never heard of them. A recurring theme in many plant families is that their tropical members are trees that do not survive in winter-cold temperate gardens. We mostly grow only the cold-tolerant perennial members of these families and seldom their woody relatives.

How much deeper one's appreciation of plants becomes as these patterns are revealed. Perhaps more fascinating than the primula–shooting star–cyclamen example is that jasmines and lilacs are olives, or at least all belong in the olive family, Oleaceae. Once again, look at the flowers: jasmines, lilacs, and olives all have four petals partly united in a tube or cup, with the free portions spreading above, and they all have just two stamens (the male portion of the flower). Also, the fruits of jasmines (the genus *Jasminum*) and olives (*Olea*) are structurally similar, fleshy with a hard inner stone containing a single seed. Lilacs (*Syringa*) have just such flowers too (some doubled cultivars excepted) but quite different fruits, dry capsules that split vertically into sections and, to the dismay of many, persist untidily on the branches through the winter months. Many other plant families have flowers with petals partly united but differ in other features, for example, more stamens, or different kinds of fruits, or a more complex arrangement of the petals.

The example of the honeysuckle family Caprifoliaceae shows another aspect of an understanding of plant families and their relationships. Caprifoliaceae include not only climbing honeysuckles (*Lonicera* species) but also shrubs and small trees such as abelia (the genus *Abelia*), beauty bush (*Kolkwitzia*), and weigelia (*Weigela*) to name some. Elderberry (*Sambucus*) and viburnum (*Viburnum*), often included in Caprifoliaceae, are classified as a separate family, Viburnaceae. These share similar small flowers with five petals, five (or fewer) stamens, an inferior ovary (the calyx and corolla are inserted at the top of the ovary), and leaves in pairs opposite one another. In a more recent development in family classification, perennials such as scabious or pincushion flower (*Scabiosa*) and valerian (*Valeriana*) are regarded as members of the honeysuckle family. Historically, these herbaceous plants have been assigned to two other families, Dipsacaceae and Valerianaceae, respectively, but their flowers match exactly those of most other

honeysuckles, and they have opposite leaves, as do honeysuckles. The close relationship of these families is corroborated by molecular genetic studies.

From these examples, it is evident then that important criteria for recognizing plant families include the number of petals (and sepals), the union or not of the petals, the number of stamens, and the position of the ovary in relation to the petals and sepals. Other features that may be important are division of the leaves (simple, or compound and comprising several leaflets) and their arrangement on the stem (either opposite or alternate, or more correctly in spirals, or sometimes in whorls). Fruits may also be important, but often less so. In the rose family (Rosaceae) fruits may be fleshy drupes (peaches, cherries), berries (raspberries), pomes (apples, pears), or dry capsules (as in spiraeas).

A passing knowledge of plant parts, especially those of flowers, is valuable in the beginning to unravel the intricacies of family classification, but it is not essential. The Glossary explains the meaning of technical terms we use, and there are surprisingly few. We hope, then, that this book will impart a deeper appreciation of plants and perhaps give you the ability, when meeting a new plant, to determine from a set of simple criteria to which family it belongs.

Family Classification

The formal classification of plants into families is relatively recent in the history of science. Efforts to place genera in what were perceived to be useful groupings were first simply aids to identification. It was the innovative Swedish botanist Carl Linnaeus who famously grouped the plants known to him according to the number of stamens (male parts of the flower) and then the number of carpels (female parts). Thus all plants with three stamens and one apparent carpel were classified as *Triandria* (three males) *monogynia* (a single female), irrespective of other features. This simple system, with its titillating sexual imagery, was an immediate success upon its publication in *Species*

Plantarum in 1753. Its limitations, however, became apparent as more and more plants became known to science as the globe was explored in the 18th century and later. Devised almost exclusively from the plants of Europe and North America, Linnaeus's system became not only unwieldy but lacked predictability. Plants evidently closely related sometimes fell in quite different clusters. True relationships, or what appeared to be so using the methods of the time, seemed an essential aspect of classification. It may seem self-evident today that related plants should be placed in the same family or genus, but that was only gradually appreciated at that distant time.

The taxonomic rank of family was introduced by the French scientist Pierre Magnol (for whom the genus *Magnolia* was named) as long ago as 1689. The word *family* in English gradually replaced the Latin word *ordo* for what was more or less the same rank. A second French botanist, Antoine Laurent de Jussieu, provided what became the baseline for family classification in 1789 in his seminal work, *Genera Plantarum*. He described for the first time many families that we recognize today and accepted others already named. (Nomenclatural priority for family names thus dates from 1789.) Time has, of course, resulted in the refinement and expansion of Jussieu's system of families.

The relationships of certain genera and species were not, and still are not, always self-evident. In addition, specialists in plant family classification differed in what they considered to be significant characters for classification. The details of older classification systems are largely of historical interest today because molecular studies, in particular the sequences of the components of DNA, the genetic material of all life, provide powerful evidence about relationships. In the case of the flowering plants, a consortium of scientists using the name Angiosperm Phylogeny Group combined this information, published in thousands of technical papers, into a refined and more accurate classification. DNA sequences provide objective, often unambiguous evidence about the relationships of genera and thus show how genera may best be arranged into families.

Plant Nomenclature

Intimidating to the uninitiated, plant nomenclature seems at first like a foreign language, which in fact it is, being derived from Latin and classical Greek. Plant nomenclature is internationally recognized and ordered so that it provides a uniform system of names across the world and its languages. Common names, in contrast, by their very nature differ for the same plant from one country to another and often even locally within a country. Moreover, the same common name is often used in different parts of the world for different plants. The need for a uniform system of plant names is thus self-evident.

Botanists and zoologists use the binomial system of biological nomenclature for species names. This system recognizes the genus as the most important rank in classification, applied as the group name for one or more species that share several common features. Scientifically, plant species names are always binomials, comprising two elements, for example, *Acer* (the generic name) and *palmatum* (the specific name) for the species we know as Japanese maple. All species of the maple genus *Acer* share a number of distinctive and sometimes unique features, most notably their two-winged fruits and small flowers with four sepals, four petals, and eight stamens. Because generic names are regarded as Latin (or Greek), they have what is called for convenience grammatical gender. Names may be masculine (most often ending in *-us*), feminine (often ending in *-a*), or neuter (ending in *-um*). When used as adjectives, species epithets must agree in gender with the genus. Thus, for example, the endings of specific names usually agree with the ending of the genus, as in *Acanthus spinosus*, *Tellima grandiflora*, or *Alyssum montanum*. The gender of Greek names is unfortunately less simple to determine, but species names follow the same pattern as those derived from Latin, agreeing with gender.

Species names can be formed in ways other than using an adjective. Names indicating the country or region of origin may end in *-ensis* if no classical adjective is available, as in *Lysichiton camtschatcensis* (from the Kamchatka Peninsula), or may be genitive, thus *Nemophila menziesii*, named after the Scottish plant explorer Archibald Menzies. If intended to honor someone, perhaps the discoverer of the species or the source of funding for an expedition, the ending *-ana* can be used, as in *Pinus banksiana* (the eastern North American jack pine), the latter named for Joseph Banks, the wealthy patron of botany and plant exploration.

There are exceptions to these rules, as there always seem to be. Perhaps the most bizarre exception is that classical trees, those known in antiquity, are usually treated as feminine despite the formation of their generic name. So *Quercus* (the oak genus, with an apparent masculine ending) is treated as feminine, thus *Quercus coccifera*, Kermes oak.

This outline should help explain the complexity surrounding plant nomenclature. Whole books have been devoted to the subject and may be consulted for additional information. To find out the meaning of plant names, the books by Stearn (1992) and Coombes (2012) are valuable resources and make fascinating reading.

Related genera are grouped into families. Family names are formed from the name of a particular genus in that family plus the ending *-aceae*, thus Primulaceae (from the genus *Primula*), Rosaceae (from *Rosa*), and so on. Plant family names are treated grammatically as plural (thus "the Rosaceae are"). In scientific and horticultural writing, generic and specific names are rendered in italics but family names are not (nor are names of subgroups of families: subfamilies, tribes, etc.). Names of genera are capitalized, but not those of species, even when named after a person or country. Families may be subdivided into subfamilies, in which the ending *-oideae* informs the reader of the rank. Thus, for example, the bean family Fabaceae are currently divided into Caesalpinioideae, Mimosoideae, and Faboideae (sometimes called Papilionoideae). Again, the name of a subfamily is properly based on the name of one of its constituent genera. Large families or subfamilies may be further subdivided into tribes and subtribes.

The *International Code of Botanical Nomenclature* governs the details of plant nomenclature, including the formation of names. One of its primary concerns is priority of publication, which determines the accepted name of a genus, species, or even family. The use of the earliest validly published name for a genus, species, or family is mandatory unless, for various reasons, a later name is formally conserved by a designated committee. Changes in well-established plant names are sometimes due to the discovery in older literature (beginning with the year 1753, when the binomial system of naming plants was established) of a prior name for a known species. Generic names may change for the same reason or because new information about relationships is discovered that makes current nomenclature inappropriate. Thus a genus may be split into one or more segregate genera. Or two or more genera may be united because the characteristics that were used to separate them are no longer significant in defining that genus. When genera are united, the combined genus takes the name of the one that has historical priority.

The *Code* allows for just a handful of alternatives in family names of archaic and historic usage. The grass family, mostly called Poaceae following current rules for forming family names, may also be called by its historical name, Gramineae. The pea or legume family Fabaceae may be called Leguminosae. The remaining alternatives are Umbelliferae for Apiaceae (carrot, celery, or parsley family), Palmae for Arecaceae (palm family), Compositae for Asteraceae (aster or daisy family), Cruciferae for Brassicaceae (mustard family), and Labiatae for Lamiaceae (mint or sage family). One remaining alternative name is Guttiferae for Clusiaceae (allied to Hypericaceae, which is described following Violaceae in Families A–Z). Currently, the use of either name for these families is acceptable, but the older, historical names are fading from the scientific literature.

Another major reason for changes in family, genus, or species names or their circumscription (loosely, their definition) is because new information renders current understanding out of date. For example, some important diagnostic structures may be reinterpreted by microscopic examination or chemical analysis. Most current changes to family classification result from comparison of the sequences in genes of the four molecules that make up the genetic molecule DNA. These sequences are understood to represent powerful information about relationships of plants at all ranks. The philosophy underpinning classification is that genera and families must be inclusive. Thus all species sharing a set of characteristics, including those provided by DNA sequences, must be treated in the same genus. Likewise, species sharing characteristics of another genus must be placed in that other genus. One of the more striking consequences of such studies is that the genus *Fragaria* (strawberries) has overwhelming molecular similarities to the larger genus *Potentilla* (cinquefoils). Strawberries are in fact potentillas with fleshy fruits, sharing similar flowers, leaves, and genetic signatures. There are hundreds of other comparable examples at generic and family rank, and those that affect temperate horticulture are outlined in the main chapter of this book, Families A–Z, which presents details of families important in temperate horticulture.

One more word about classification: Families are grouped into higher-ranking groups, including orders, superorders, and classes, mainly for scientific purposes. Flowering plants comprise just one, admittedly large and attractive group among the land plants. Earlier classifications recognized three main groups of land plants, namely, flowering plants (class Magnoliopsida), gymnosperms, and the ferns and fern allies. Modern classifications recognize that the situation is far more complex, with higher levels of diversity among the nonflowering plants than previously appreciated. Flowering plants are now classified as the subclass Magnoliidae, equivalent in status to the conifers, cycads, maidenhair trees (*Ginkgo*), and jointfirs (*Gnetum*).

There are three major groups of flowering plants. The paleodicots or ancient dicots include *Magnolia* and a range of related, ancient flowering plant families that have distinctive pollen grains, unspecialized water-conducting tissues, and a mostly undefined number and arrangement of floral organs. The monocots are the lily- and grass-like families, plus the palms, all of which have distinctive leaves and leaf venation, more specialized water-conducting tissues, and an established number of floral parts, most frequently in multiples of three. Lastly, the eudicots or modern dicots include families with specialized pollen grains and water-conducting tissues, and floral parts in set numbers, most often multiples of five, less often four or even three, or other numbers in particularly specialized families of genera.

The flowering plants include by far the greatest number of land plants, an estimated 250,000 to 300,000 species. New species are constantly being discovered and named, and known species may be reappraised and split up into two or even more species. The estimated 300,000 species are assigned to as many as 450 families, with consensus on the number and circumscription still not settled. Given this large number it is perhaps surprising that plants cultivated for food, medicine, ornament, timber, and other purposes belong to no more than 150 families, and less than half of these include significant numbers of genera and species.

We treat in detail 92 families, listed alphabetically in the main section of the book, Families A–Z. In addition, 35 more families are described with or following related or similar families in the A–Z. (There is an alphabetical list of these 35 families at the beginning of the A–Z.) Finally, there are 30 additional families listed alphabetically by genus in the appendix, Genera of Small Families Otherwise Not in General Cultivation. Thus 157 families of temperate garden plants are described in this book. The Index includes not only the scientific names of the families, genera, and species but also many cross-references by common names.

Families are illustrated in detail with line drawings, and many are supplemented with color photographs, most taken in their wild habitats. Used in conjunction with the line drawings, the photographs show some of the variation in the larger families.

We decided not to include one traditional feature of a book of this kind, a key to families. A key consists of a series of paired questions that ultimately lead to the identity of the family to which the plant in question belongs. Such keys often end up utilizing characters that are hard to observe and that require more technical knowledge and terminology. The next chapter is intended as a brief guide to plant structure.

Despite conflicting information about some of the more recent changes in family circumscriptions, the ones affecting the families treated here are relatively few. Most significant are the following:

- Alliaceae (onion family) and Agapanthaceae are often included in Amaryllidaceae; for convenience, we deal with Alliaceae as a separate family.

- Asclepiadaceae (milkweed family) are now a subfamily of Apocynaceae, the Asclepiadoideae.

- Asparagaceae (asparagus or agave family) include several small families sometimes separately recognized, including Agavaceae, Convallariaceae (lily-of-the-valley family), and Hostaceae. The relatively large Hyacinthaceae have also been placed in Asparagaceae, but we deal with that family alphabetically in the Families A–Z because it includes so many cultivated plants.

- The genus *Calceolaria* has bounced from Scrophulariaceae (figwort family) to the mostly tropical family Gesneriaceae (described following Bignoniaceae) but now enjoys separate familial status as Calceolariaceae (also described following Bignoniaceae).

- Campanulaceae (bellflowers and relatives) include genera of the former Lobeliaceae, which comes as no surprise.

- Caprifoliaceae (honeysuckle family), mostly woody plants, include the herbaceous families Dipsacaceae and Valerianaceae, but *Sambucus* and *Viburnum* have been assigned to Viburnaceae (elderberry family). Viburnaceae are discussed following Caprifoliaceae.

- Chenopodiaceae are subsumed into Amaranthaceae.

- Opinion is divided regarding treatment of Sapindaceae (soapberry family, which may include Aceraceae, maple family) and Hippocastanaceae (horse chestnut family); we deal with all three separately as befits their horticultural importance and as favored by several experts. Aceraceae appear alphabetically in Families A–Z; Sapindaceae are described following Hippocastanaceae.

- Several genera of Scrophulariaceae (figwort family) have been raised to separate family status as Phrymaceae (including *Diplacus* and *Mimulus*) and Plantaginaceae (with *Antirrhinum, Bacopa, Penstemon, Veronica*, etc.). Because of their similarities to Scrophulariaceae, Phrymaceae and Plantaginaceae are discussed together with Scrophulariaceae.

- Sterculiaceae are now included in Malvaceae (mallow or hibiscus family).

Plant Morphology

Here we discuss the features of plant form that are most useful in identifying plant families. More information on plant morphology may be found in the extensively illustrated book by Bell (2008).

LEAF The structure and arrangement of leaves are important in understanding plant structure and their classification. Thus a grasp of leaf terminology is essential for recognizing or identifying plant families and genera, many of which are consistent for certain leaf characteristics. Leaves may be inserted on the stem in spirals (often termed alternate), in pairs opposing one another (two leaves thus exactly opposite), or in whorls of four (or more) arising at the same level on the stem. In many annuals and herbaceous perennials the leaves are crowded at the base of a plant, and then their arrangement on the stem is simply not evident. Leaf insertion is frequently, although not always, consistent within a family or genus. Thus all maples (*Acer*) have opposite leaves whereas all members of Fagaceae (oak or beech family) have leaves in spirals. Exceptions to this pattern can occasionally occur in large families. Thus, while members of Rosaceae (rose family) typically have leaves in spirals, a few genera have opposite leaves, as in *Lyonothamnus* and *Rhodotypos*, which otherwise have the morphological and molecular markers of the family.

Leaves consist of a stalk or petiole, and a flattened blade or lamina, and may be simple (with one undivided but often variously lobed blade) or compound (with multiple separate blades or leaflets). Leaves without a petiole are termed sessile (without a stalk). In nearly all plants there is a bud (or lateral branch) on the stem in the angle (called an axil) formed by the petiole and stem. The point on the stem where the axillary bud and subtending leaf are located is termed a node, and the length of stem between nodes is an internode. It may be difficult to distinguish a single compound leaf from a shoot bearing several simple leaves. Compound leaves have a bud at the base of the main axis or stem but not at the base of each leaflet. Some simple leaves can be deeply lobed, superficially resembling compound leaves, for example the genus *Cannabis*, marijuana.

Compound leaves may be pinnate (like a feather), with the leaflets (or folioles) arranged in two opposed lines along the main axis, or they may have only three leaflets, thus trifoliolate. Leaves in which several leaflets radiate from the apex are termed digitately compound, a feature characteristic of, for example, the buckeyes, the genus *Aesculus*. Pinnately compound leaves with a terminal leaflet are termed imparipinnate, whereas those lacking a terminal leaflet are termed paripinnate (as in most genera of Sapindaceae, soapberry family, described following Hippocastanaceae), a somewhat unusual condition.

The shape or outline of simple leaves is sometimes diagnostic for genera or even families and may typically be described as entire (without lobes), pinnately lobed (as in a feather), or palmately lobed (as in a hand). Leaves of many maples, for instance, are palmate. Leaf margins may be smooth or variously incised or toothed. Toothed margins with symmetric teeth are termed dentate, and those with asymmetric teeth or saw-toothed are termed serrate. Leaves with margins shallowly and evenly lobed are described as crenate.

Families are often consistent in leaf form. Entire leaves are present in the majority of families, so compound leaves are thus helpful in identifying particular ones. Families in which compound leaves are the ancestral but not necessarily the only condition include Bignoniaceae, Fabaceae, Juglandaceae, Rosaceae, and Sapindaceae, to name some prominent examples.

INFLORESCENCE A rather intricate terminology has been developed to describe the range of different ways in which flowers are presented. Typically, flowers are either solitary or arranged in clusters called inflorescences, which may terminate a branch or form axillary clusters (sometimes called thyrses). Indeterminate inflorescences are those in which the lowermost flowers open first and are called racemes unless the individual flowers are sessile (without individual stalks) and then are spikes. Determinate inflorescences are those in which the terminal flower opens first, followed by those lower down; these are termed cymes. For purposes of this book little more detail is necessary. Multibranched inflorescences are loosely called panicles. Flowers may also be grouped in tight heads, for which the term capitate is often used, but if the individual flowers are each stalked then the inflorescence is called a corymb, unless, as in the Apiaceae (carrot family), all arise from the same point and are raised to the same level, thus more or less flat-topped, in which case the inflorescence is known as an umbel. These distinctions are adequate for most purposes. The use of some technical terms saves using multiple words repeatedly to describe the same inflorescence.

FLOWER The structure of flowers, more than any other feature, characterizes families. Floral structure includes, in a sequence from outer to inner, all or some of the following components: the perianth, consisting of sterile leafy or colored organs surrounding or enclosing the fertile parts; the male parts, the stamens (collectively the androecium from the Greek, male house); and the female parts, individual units of which are called carpels (gynoecium, female house). The floral parts are attached to the receptacle, the tissue at the apex of a flower stalk, which is called a pedicel. When male and female organs are present in a flower, the most common state, flowers are termed perfect or bisexual. Any one of the three sets of organs may be absent (for example, in unisexual male or female flowers, one or the other of the fertile whorls is absent or sterile).

The individual elements or parts of the perianth may be spirally arranged or in concentric rings or whorls, typically in just two whorls. When the two whorls are alike in color and texture the individual elements are usually called tepals. In many instances, however, the two whorls differ markedly from each other and in such flowers the outer whorl is the calyx, with the individual sepals often green and somewhat leaf-like, whereas the inner whorl or corolla consists of colored petals. The units of one or both whorls may be partly fused together (connate or united), at least basally, or joined into a tube, cup, or funnel. It is useful to note that this condition is consistent in families. Thus sets of families with united petals were historically united in a rank above that of family as the Sympetalae. In the same vein, families with multiple tepals were united as the Polypetalae. The Sympetalae are now understood to be an artificial group with the union of petals occurring multiple times during the course of evolution. Thus families in which petals are partly united are not necessarily related.

The number of tepals (perianth lobes), sepals (calyx lobes), or petals (corolla lobes) may be indefinite or may be consistent in a fixed number, often six in the monocots but most frequently four or five in the dicots. Two or more flower whorls often share the same number of units, and for convenience we use a shorthand, calling flowers with parts in multiples of three, three-merous, those with parts in multiples of four, four-merous, etc., thus implying the number of sepals, petals, often the stamens and sometimes also the carpels.

The male parts of the flower, the stamens, may likewise be arranged either in spirals of many or whorls of a set number. Typically, each stamen consists of a stalk, the filament, which is often but not invariably thread-like, and a terminal organ, the anther. The anther consists of two discrete lobes, each comprising a pair of cells or thecae that contain the pollen grains. The anther lobes are embedded in sterile tissue called the connective. When mature, usually after a flower bud has opened, the anther cells split along a predetermined line, usually along longitudinal slits but sometimes by pores or in a few families opening by terminal flaps (called valves, for example, Lauraceae, the laurel or bay family). Anthers may face inward toward the center of the flower, termed introrse, or outward toward the exterior, termed extrorse. In a few families anthers may be latrorse, thus opening by lateral slits located on the sides of the anthers.

Female parts of the flower, the carpels, may also be arranged in spirals of many or whorls of a set number. The carpel consists of an ovary, containing one or more ovules, and a terminal stalk or style, terminating in a stigma, the surface on which pollen is deposited to begin the process of fertilization (union of the male and female gametes). Carpels are frequently closely united so that the number of units may not be immediately evident without dissection. The number of styles or stigmas may provide a clue since these often reflect the number of carpels even when those are otherwise completely joined or fused. In such cases the only reliable way of determining the number of carpels is to cut across the ovary to expose the cells of the constituent carpels. Another significant aspect of the carpels is that the ovary may be superior, with perianth parts inserted below the ovary, or inferior, with perianth parts arising above the ovary. In the latter case the ovary (not the style and stigma) is then not visible looking inside the flower. Most families are consistent for ovary position, with the inferior ovary considered a specialized feature.

FRUIT One of several unique features of flowering plants is the development of a fruit that encloses the seeds during their maturation. The word *angiosperm*, often used for a flowering plant, simply means fruit with seeds. A complex terminology has been developed to describe fruits. Fruits are derived by growth of the ovary wall after fertilization of the ovules. Mature fruits are typically either dry or fleshy. Dry fruits derived from a compound ovary (of multiple fused carpels) and containing multiple seeds are called capsules. Those from separate carpels are follicles. Fleshy fruits containing a single seed are called drupes (for example, peaches and plums), and those containing multiple seeds are called berries, a term more precisely defined than our language generally allows. Somewhat confusingly, many fleshy fruits are derived not from the ovary proper but from surrounding tissues. In strawberries, for instance, the receptacle is fleshy and edible and the individual fruits are those hard, dry grains on the surface. A raspberry is by definition a cluster of drupes. Nuts are dry fruits containing a single seed and a hard fruit coat. A fig fruit develops from an entire inflorescence containing multiple flowers, each flower forming after fertilization a fleshy one-seeded fruit but together forming a compound structure (an infructescence, in the case of figs called a syconium). A passing familiarity with fruit terminology will aid in understanding family descriptions.

FAMILIES A-Z

There are 92 families listed alphabetically by scientific name in the A–Z. An additional 35 families are described with or following related or simillar families. To facilitate locating those, they are listed alphabetically here along with the family each follows. Finally, a further 30 families are described in the appendix, Genera of Small Families Otherwise Not in General Cultivation. The Index has many cross-references by common names to plant families, genera, and species.

Families Described in Passing in the A–Z

Alstroemeriaceae follow Liliaceae

Aquifoliaceae follow Celastraceae

Asteliaceae follow Asparagaceae

Calceolariaceae follow Bignoniaceae

Calycanthaceae follow Magnoliaceae

Capparaceae follow Brassicaceae

Cleomaceae follow Brassicaceae

Cucurbitaceae follow Begoniaceae

Dasypogonaceae follow Asphodelaceae

Ebenaceae follow Styracaceae

Garryaceae follow Cornaceae

Gesneriaceae follow Bignoniaceae

Hemerocallidaceae follow Asphodelaceae

Hypericaceae follow Violaceae

Iteaceae follow Hamamelidaceae

Juncaceae are with Poaceae

Linaceae follow Violaceae

Loganiaceae follow Gentianaceae

Melianthaceae follow Geraniaceae

Myricaceae follow Fagaceae

Nelumbonaceae follow Nymphaeaceae

Nothofagaceae follow Fagaceae

Paeoniaceae follow Ranunculaceae

Phrymaceae are with Scrophulariaceae

Phyllanthaceae follow Euphorbiaceae

Phytolaccaceae follow Nyctaginaceae

Plantaginaceae are with Scrophulariaceae

Restionaceae are with Poaceae

Sapindaceae follow Hippocastanaceae

Simmondsiaceae follow Buxaceae

Strelitziaceae follow Zingiberaceae

Ulmaceae follow Cannabaceae

Viburnaceae follow Caprifoliaceae

Xanthocerataceae follow Hippocastanaceae

Xanthorrhoeaceae follow Asphodelaceae

Acanthaceae, *Asystasia gangetica*. The zygomorphic flowers, either white or blue, have five petals united below in a trumpet-shaped tube with spreading petal limbs that enclose the superior ovary with slender style and two-lobed stigma. The four stamens are in two pairs with anthers that, as in many other genera of the family, are shortly tailed or spurred. Typical of Acanthaceae are its two-parted, explosive capsules.

ACANTHACEAE

Acanthus Family

About 3800 species in 202 genera

RANGE Nearly cosmopolitan, especially dry tropics, few in cold temperate zones

PLANT FORM Shrubs, a few trees, vines, some perennial herbs, often with swollen nodes; leaves simple, opposite or basal (*Acanthus*), often decussate, entire to deeply dissected (but not compound), without stipules, hairless or glandular haired

FLOWERS Usually zygomorphic and usually 2-lipped, perfect, subtended by bracts that are often colored and prominent or sometimes spiny, borne in bracteate racemes or spikes; calyx usually of 5 (or 4 or 6) sepals united below, lobes often sharply pointed; corolla petaloid, usually with 5 or 6 lobes (3 in *Acanthus*) united below in a short or long tube; stamens 2, 4, or 5, inserted on the corolla tube, anthers often tailed or crested; ovary superior, compound, of 2 united carpels and 2 cells with a single terminal 2-lobed style

FRUIT Usually a dry capsule, more or less club-shaped, with two to several seeds shed explosively

Although a relatively large family, Acanthaceae provide relatively few subjects for temperate horticulture, all of them in two of the four subfamilies that are recognized, Acanthoideae and Thunbergioideae. The latter includes several extremely attractive vines, some shrubs, and a few perennials of the genus *Thunbergia*, hardy in warm temperate and tropical areas. The southern European and western Asian *Acanthus* includes several species that thrive in temperate gardens, especially *A. mollis*, *A. spinosus*, and *A. syriacus*, the last a dwarf species with interesting lobed and spiny leaves, ideal for rock gardens. The leaves of *A. mollis* or more likely *A. spinosus* reputedly provided the inspiration for the leaf motif adorning the capitals of Corinthian columns. The common name for *Acanthus* species, bear's breeches, appears to be a corruption of the older common name bear's claws, which better fits the flowers with their spiny calyx lobes.

Genera widely grown in warmer climates include *Crossandra*, *Justicia* (including *Beloperone*), *Hypoestes*, and *Ruellia*. The shrimp plant (*J. brandegeeana*), native to Mexico, is unfortunately not hardy and so must be overwintered indoors except in warm temperate and Mediterranean climates. The shrub *Mackaya bella*, from southern Africa, has particularly fine, large flowers of the palest mauve and makes a handsome hedge in areas with a warm temperate climate. Species of *Strobilanthes* are grown in greenhouses or as summer annuals, notably the purple-leaved Persian shield, *S. dyerianus*. At least *S. penstemonoides* is hardy and an interesting garden subject for light shade. Unfortunately, the attractive dark blue flowers often fail to open in hot weather. The family also includes the mangrove tree (growing in tidal saline water along coasts) *Avicennia*.

Acanthaceae, *Mackaya bella*

Acanthaceae are recognizable by the opposite and often decussate leaves (clustered at the base in some species), always simple and without stipules, zygomorphic flowers with the corolla lobes united in a short to long tube, and mostly four or five stamens. The ovary is superior and two-celled, and the terminal style has two stigmatic lobes. The capsules, containing few, relatively large seeds, are distinctive though seldom seen in garden situations. The seeds are shed explosively as the capsules split open.

Acanthaceae provide no plants of significant medicinal or agricultural importance. Several species are used locally as green vegetables and as medications. The Asian *Asystasia gangetica* (Chinese violet, creeping foxglove) is weedy across tropical parts of the world.

Selected Genera of Acanthaceae
Acanthus • Asystasia • Barleria • Blepharis • Brilliantaisia • Crossandra • Dicliptera • Hypoestes • Justicia, including *Beloperone* and *Monechma • Mackaya • Ruellia • Strobilanthes • Thunbergia*

Acanthaceae, *Acanthus syriacus*

ACERACEAE

Maple Family

About 127 species in 1 genus

RANGE Mainly northern temperate, a few in tropical mountains, many in eastern Asia

PLANT FORM Large and small trees, shrubs; leaves opposite, simple, often palmately lobed or compound and trifoliolate or pinnate; without stipules

FLOWERS Bisexual or unisexual, mostly in cymes or racemes, small, radially symmetric; sepals free, mostly 4 or 5, usually colored; corolla with as many petals as sepals (or petals lacking), free; stamens 3–12, mostly 8, anthers with longitudinal slits; ovary superior, of 2 united carpels, 2-lobed and 2-celled, usually with 1 or 2 ovules in each

FRUIT A distinctive two-winged and two-seeded winged structure, a samara, rarely the wing circumferential, dispersed by wind

Aceraceae comprise a single genus of woody plants, *Acer*, most of the species deciduous. Maples are familiar to most gardeners in the temperate zone, and many gardens include at least one maple. Species are grown primarily for their shape, moderate size, and in many, striking autumn color. Dozens of cultivars are available, mostly of the so-called Japanese maple, *A. palmatum*, but also of other species. Cultivars include those with dwarf habit, lacy leaves, and leaf color variation, not only those with pale or deep red pigment but pale green to almost white with green veins, variegated with white margins, and many more. Several cultivars of *A. japonicum* (full moon maple) are also grown in temperate gardens for their brilliant autumn color. The western North American *A. circinnatum* (vine maple) is more closely related to the Asian *A. palmatum* group than to other American maples. *Acer negundo*, with compound leaves, is another North American native that is widely grown and has become invasive in Europe. Several *Acer* species are tall trees used in urban forestry or for timber. Some are reputed to have a sweet-tasting bark, and the sugar maple, *A. saccharum*, has sugary sap, which is tapped in spring when growth begins. After boiling to concentrate the sap, the product is marketed as maple syrup. The genus *Dipteronia*, distinctive in having a seed with a circumferential wing, is now included in *Acer*. Fossils with this fruit are recorded from Eocene deposits (over 40 million years old) in northern Canada.

Acer has also been included in the Sapindaceae (soapberry family), which have leaves in spirals, together with *Aesculus* (buckeyes and horse chestnuts). Botanists are divided on this issue, and here we regard Aceraceae and Hippocastanaceae (the family to which *Aesculus* belongs) as separate from Sapindaceae. There is more extended discussion of Sapindaceae following Hippocastanaceae.

Aceraceae are recognizable primarily by the fruit, the familiar two-winged, two-seeded samara (each seed with a single wing), also by the small, often inconspicuous flowers with colored sepals and superior ovary. Leaves are opposite and range from simple, then often with palmate venation, to compound, then digitate, trifoliolate, or pinnate.

Genera of Aceraceae
Acer, including *Dipteronia*

Aceraceae, *Acer macrophyllum*

Aceraceae, *Acer palmatum*

AGAPANTHACEAE

Agapanthus Family

7 species in 1 genus

Agapanthus has many characteristics of Amaryllidaceae but differs primarily in having a superior rather than inferior ovary. The rootstock is a fleshy rhizome rather than a bulb, but some Amaryllidaceae also have a fleshy rhizome, notably *Clivia*. The flowers, always in umbels, are shades of blue to violet (or white), a color rare in Amaryllidaceae (the "blue amaryllis" *Worsleya* has lilac to light blue flowers), and the capsules reflex sharply at maturity, giving the fruiting heads a characteristic look. *Agapanthus*, with about 12 species, is restricted to southern Africa. The genus is widely cultivated in areas of mild to tropical climate. Two deciduous species, *A. campanulatus* and *A. inapertus*, are more tolerant of areas with cool temperate climates. Many cultivars are now available in the trade. The common name for the genus is either "agapanthus" or that misnomer, "lily of the Nile."

Genera of Agapanthaceae

Agapanthus

Agapanthaceae, *Agapanthus caulescens*

Aizoaceae, *Carpobrotus deliciosus*. As in many Aizoaceae, the succulent leaves are opposite, with each pair shortly united at the base. The inferior ovary, topped with several plumose stigmas, matures into an edible, fleshy fruit that is characteristic of the genus and from which it receives its common name. The numerous filamentous petals are actually modified sterile stamens, colored with water-soluble red pigments.

AIZOACEAE

Ice Plant Family

About 1850 species in 125 genera

RANGE Dry tropics and warm temperate climates but overwhelmingly southern African, especially in the winter-rainfall zone

PLANT FORM Shrubs, subshrubs, perennial herbs (many trailing or prostrate), some annuals; leaves simple, mostly opposite or basal, succulent (but stems woody or weakly succulent), entire, often round in cross section rather than bifacial, occasionally lobed, rarely with stipules

FLOWERS Usually bisexual, solitary or in cymes, often strongly diurnal; sepals free, 4 or 5, sometimes brightly colored; stamens 4 or 5 to many, then outer ones often sterile and flattened into narrow, glossy, petal-like organs in as many as 6 whorls (flowers often superficially resembling daisies), inner sterile and fertile stamens sometimes aggregated in a cone around styles; ovary superior or inferior, of 2–5 or many united carpels with as many cells or 1-celled, usually with many ovules in each

FRUIT Dry capsules or rarely fleshy fruits, many opening only when moistened, seeds then dispersed by raindrops

Almost exclusively succulent, Aizoaceae include several genera of horticultural importance and potentially many more. Some species have raised, bladder-like cells on the leaf surface that glisten like ice crystals, giving rise to the common name, the ice plant family. Many species tolerate light frost but few survive temperatures more than a few degrees below freezing, limiting their use in temperate gardens. Nevertheless, several species of *Delosperma* from high elevations are useful ground covers, and some have stunningly colored flowers. In gardens in areas of warm temperate and Mediterranean climates, species of *Drosanthemum*, *Lampranthus*, and *Ruschia* make brilliant displays in spring, the glossy, white, pale to deep pink, purple, or red flowers covering the sprawling plants. The shrubby plants are short-lived, but stem cuttings are easily rooted to replace old specimens. Many genera are valued garden plants in dry parts of the world. What look like petals are actually petal-like sterile stamens; those of the largest and most diverse group, subfamily Ruschioideae, are particularly shiny and reflective. Annuals of the genus *Cleretum* (formerly *Dorotheanthus*) are particularly striking spring ephemerals and are widely grown in areas of Mediterranean climate or those with mild winters, especially in southern Africa.

Aizoaceae are also widely grown indoors by succulent enthusiasts, and dwarf species of many genera are maintained under glass, perhaps most notably the dwarf succulents *Argyroderma* and *Gibbaeum* as well as the stone plants *Conophytum* and *Lithops*. Those plants produce just two new leaves a year, each plant thus resembling

Aizoaceae, *Drosanthemum speciosum*

Aizoaceae, *Carpobrotus quadrifidus*

Aizoaceae, *Conicosia elongata*

Aizoaceae, *Cleretum (Dorotheanthus) maughanii*

a small stone, with the color and patterning often mimicking the pebbles among which they grow in the wild.

Species of the southern African genus *Carpobrotus* are used in street plantings around the world, and in coastal areas to reduce erosion, sometimes becoming weedy, for example, *C. deliciosus* (sour fig). The fleshy fruits of *Carpobrotus* are used locally for jams and condiments. One species of *Tetragonia*, a genus of southern Africa, South America, and Australasia, *T. tetragonioides* is used as a potherb, now grown around the world as New Zealand spinach. Plants were introduced to Europe from New Zealand by

Captain Cook in the 18th century and were highly valued for their relief of scurvy.

Selected Genera of Aizoaceae

Aizoon • Argyroderma • Carpobrotus • Conophytum • Cleretum, including *Dorotheanthus • Delosperma • Drosanthemum • Faucaria • Galenia • Gibbaeum • Lampranthus • Lithops • Mesembryanthemum • Oscularia • Ruschia • Tetragonia*

ALLIACEAE

Onion Family

About 850 species in 33 genera

RANGE Nearly cosmopolitan, not Australia, with centers of distribution in Eurasia and South America

PLANT FORM Perennial herbs with bulbs, corms, or rhizomes; leaves spirally arranged, arising from the bulb; stems leafless, often fleshy, flat or round and hollow, without stipules; usually more or less aromatic, smelling of onion or garlic due to sulfide compounds

FLOWERS Perfect, usually radially symmetric, borne in leafless, bracteate umbels (rarely a spike) with individual flowers stalked or sometimes sessile; calyx and corolla petaloid and usually similar, perianth thus in 2 whorls of 6 tepals, nearly free or united in a tube, sometimes with a corona (as in *Tulbaghia*); stamens 6, inserted at bases of tepals or in a corolla tube; ovary compound, of 3 united carpels with 1 terminal style, ovary superior, 3-celled

FRUIT A dry capsule with seeds angled to flattened, with shiny black coat

Although a relatively large family, Alliaceae include the huge genus *Allium*, which includes some 750 species. The family is of modest significance in horticulture but provides several important food plants, including garlic (*A. sativum*), leeks (*A. porrum*), and onions (*A. cepa*). The name shallot is derived from Ashkelon in the eastern Mediterranean, whence plants were introduced to Europe by the Crusaders, its original species name *A. ascalonicum* reflecting its source. The word scallion has the same root, but the name is an appropriation and applied to several species of *Allium* with hollow green leaves, small bulbs, and mild flavor. Chives, rampions or ramps, shallots, and several more are used locally as foods and flavorings. Other species are grown as ornamentals, many of them cold hardy. Particularly striking is the tall *A. giganteum* with flowering stems to 3 feet (about 1 meter) high and inflorescences to 5 inches (about 13 cm) in diameter. *Allium triquetrum* (three-cornered garlic), with three-sided stems, has attractive white flowers in spring but an unfortunate tendency to become invasive. Unusual for the genus, *A. bulgaricum* has quite large, cup-like, drooping flowers, borne on graceful, curved pedicels.

The southern African *Tulbaghia* is receiving horticultural attention outside its native haunts, where species have been grown for years. Mauve-flowered *T. simleri* and *T. violacea* are now grown in warm temperate gardens and street plantings. The South American and largely Chilean *Leucocoryne* has several species of ornamental value and is worth more horticultural attention. Also South American, *Ipheion* has flowering stems bearing a single flower, is hardy, and is sometimes seen in gardens.

Extracts, especially of garlic, but also other *Allium* species, are valued for their medicinal properties and have been used for thousands of years. The extracts have

Alliaceae, *Allium triquetrum*. Alliaceae are recognizable by their leafless stems, bulbs, and umbellate inflorescences but they always have
a superior ovary and radially symmetric flowers. The family contains
sulfide compounds imparting the onion- or garlic-like smell.

Alliaceae, *Allium israeliticum*

Alliaceae, *Tulbaghia acutiloba*

Alliaceae, *Tulbaghia natalensis*

antifungal and bactericidal properties and have been used with success to reduce infections, including the common cold. Allicin, one of the compounds in garlic, is an effective antibiotic and also used to reduce tumors. Gardeners note, allicin is an effective slug and snail poison. The use of garlic to ward off vampires and trolls is legendary.

Alliaceae are most easily recognizable by the umbellate inflorescence (in *Milula* a spike) of flowers with a superior ovary, rootstock with few exceptions a bulb, leaves clustered at the base, and flowering stem leafless. Floral parts are in multiples of three, and there are usually six stamens. *Agapanthus* (Agapanthaceae) is most similar but lacks the sulfide compounds that provide the distinctive smell of all Alliaceae when crushed. Amaryllidaceae share an umbellate inflorescence and bulbous rootstock with Alliaceae but have an inferior ovary and also lack the onion or garlic smell when crushed.

Selected Genera of Alliaceae
Allium • Beauverdia • Gilliesia • Ipheion • Leucocoryne • Miersia • Milula • Nectaroscordum • Nothoscordum • Tristagma • Tulbaghia

Alliaceae, *Allium roseum*

AMARANTHACEAE

Amaranth Family

About 2125 species in 173 genera

RANGE Almost cosmopolitan, mainly tropical and warm temperate, few in temperate Eurasia; particularly well represented in dry climates

PLANT FORM Shrubs, mostly small, a few small trees, many perennial herbs and annuals, sometimes with a purple or reddish tinge; leaves simple, in spirals or opposite, usually entire, sometimes succulent, without stipules; several genera are halophytes, thriving along the seashore

FLOWERS Perfect or unisexual, usually in cymes, often tiny and crowded together in axils and at branch tips, often with bristly, conspicuous bracts, these sometimes brightly colored; calyx and corolla not differentiated, 3, 4, or 5 free tepals in 1 whorl; stamens as many as tepals and opposite them; ovary superior, of 1–3 carpels, 1-celled, with 1 style and 1- to 3-lobed stigma, with several ovules or often only 1

FRUIT A nutlet, sometimes surrounded by the remains of the perianth, or several shed together as a unit

A few Amaranthaceae are useful garden plants in the temperate zone, notably *Celosia* (cockscomb), *Gomphrena* (globe amaranth), *Kochia*, and the Australian *Ptilotus*, cultivated annuals sometimes called pussy tails. Some species of *Amaranthus* are grown in gardens and have the unfriendly common name pigweed. *Alternanthera* and *Iresene* are sometimes grown in greenhouses or as annuals for their colored foliage, either dark red or multicolored. The flowers of Amaranthaceae are always small and inconspicuous, but the bracts are sometimes colored bright red, yellow, or purple and make a fair display when crowded in dense inflorescences.

Historically, Chenopodiaceae have been regarded as a separate family closely allied to Amaranthaceae, but more recent study shows it is more appropriately included therein. Thus Amaranthaceae now include many of our vegetable and salad plants. Among these, beets and chard (*Beta*) and spinach (*Spinacia*) are particularly important. Sugar beets (*B. vulgaris*), selected during the Napoleonic Wars in France for high sugar content, are grown commercially for production of sugar. Species of *Amaranthus*, *Atriplex*, and *Chenopodium* are used as potherbs, grown as a vegetable in parts of Africa. Seeds of *C. quinoa* are the grain quinoa, a crop that is becoming important as a starch source for people with intolerance for wheat and other cereal products. Originating and domesticated in Bolivia and parts of neighboring Peru, where it was a food staple, quinoa is now cultivated commercially in many parts of the world. The increased demand for quinoa in western countries has driven the price of the grain to levels at which it is sometimes no

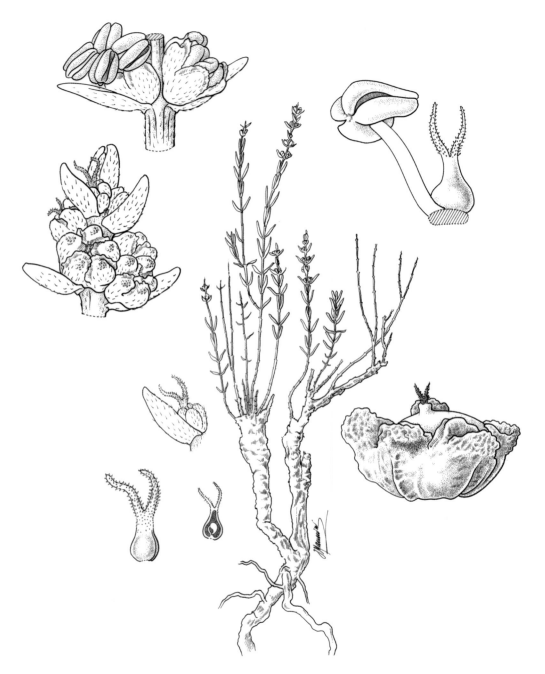

Amaranthaceae, *Chenolea convallis*. One of the succulent members of the family and formerly placed in Chenopodiaceae, chenolea is typical of the family in its simple, nearly opposite leaves without stipules, and tiny flowers without petals clustered in a spike-like inflorescence. The five undifferentiated tepals are joined at the base in a single whorl opposite the stamens, and the somewhat fleshy calyx enlarges in fruit and is shed with the single-seeded nutlet.

longer affordable in its native countries. *Chenopodium album* (lamb's quarters) is weedy in many parts of the world and a food plant in northern India. The common name goosefoot for *Chenopodium* refers to the rhombic shape of the leaves of some species.

Tumbleweeds are arid-country plants that when dry break off at the base and blow across the landscape, slowly releasing their seeds as the fruits are abraded and broken open. One of these, *Kali tragus* (or *Salsola tragus* if the genera are combined), a Eurasian native, is weedy and invasive in parts of the world, including North America and southern Africa.

Amaranthaceae are recognizable by their tiny flowers with a somewhat fleshy or papery perianth in a single whorl and usually as many stamens as tepals and inserted opposite them. The ovary is superior and usually one-celled. The prominent bristly or feathery bracts surrounding the tiny flowers of some genera, including *Amaranthus*, are distinctive and provide color to the inflorescence lacking in the tiny flowers themselves.

Selected Genera of Amaranthaceae
Alternanthera • Amaranthus • Atriplex • Beta • Celosia • Chenopodium • Gomphrena • Iresene • Kali • Kochia • Ptilotus • Pupalia • Salicornia • Salsola • Spinacia

Amaryllidaceae, *Nerine humilis*. Typical of Amaryllidaceae in its bulbous rootstock, leafless flowering stem, and umbellate inflorescence, *Nerine* has flowers with almost free perianth parts. The flowers are usually zygomorphic, with the six stamens arching downward unilaterally below the tepals. As in all Amaryllidaceae, the two whorls of the perianth are petal-like and similar, and the ovary is inferior.

AMARYLLIDACEAE

Amaryllis Family

About 900 species in 60 genera

RANGE Tropical and warm temperate, also cold temperate Eurasia, with centers of distribution in southern Africa and the Andes of South America

PLANT FORM Perennial herbs with bulbs, rarely fleshy rhizomes (*Clivia*, *Scadoxus*); stems usually more or less fleshy, leafless; leaves arising from the bulb, usually fleshy to somewhat succulent, opposite (appearing as a basal rosette), without stipules

FLOWERS Perfect, radially symmetric or zygomorphic, borne in bracteate umbels on leafless flowering stems with individual flowers stalked; calyx and corolla petaloid and similar, perianth thus in 2 whorls of 6 tepals, nearly free or variously united into a short to long tube, sometimes with a corona (*Narcissus*, *Pancratium*), symmetric or 2-lipped, usually large and brightly colored; stamens 6 (3 in *Zephyranthes*, 5 in some *Griffonia*, 18 or more in some *Gethyllis*), inserted on the corolla tube (rarely also attached to the style); ovary inferior, compound, of 3 united carpels, 3-celled, with 1 terminal style

FRUIT Usually a dry capsule, then seeds either dry with shiny black coat, or fleshy, large, green or purple, and germinating soon after falling (*Amaryllis*, *Brunsvigia*, *Nerine*); the fruits rarely berries (*Clivia*, *Haemanthus*, *Scadoxus*)

A relatively large family, Amaryllidaceae include the hugely important horticultural genus *Narcissus*, one of few cold temperate genera of the family and distinctive in the flowers having a corona, either cup-like or, as in daffodils, prominent and trumpet-like. A genus of some 50 species, *Narcissus* includes not only daffodils (*N. pseudonarcissus*) but also species with multiple flowers on each inflorescence, including *N. jonquilla* (the jonquil) and *N. tazetta* (paperwhites). Numerous hybrids and selections between different species are available. Daffodil and narcissus production for cut flowers and as bulbs for the horticultural trade is a huge industry. Hundreds of acres or hectares are devoted to their cultivation in the Netherlands, the United Kingdom, and elsewhere.

Galanthus (snowdrops) and *Leucojum* (snowflakes), native to Europe and western Asia, are also widely grown in cold temperate climes as garden subjects. Few of the many southern African and Andean genera are cold hardy, but in areas of Mediterranean climate or mild winters, *Amaryllis belladonna* (naked ladies), species of *Nerine* (for example, *N. humilis*, Cape nerine), the eastern Asian *Lycoris*, and the Andean and Mexican *Habranthus*, *Hymenocallis*, and *Zephyranthes* are useful garden plants. *Nerine bowdenii*, from the high Drakensberg of southern Africa, is hardy to about 16 degrees Fahrenheit (–9 degrees Celsius) and thus may be grown in much of North America and at least coastal Europe. Among the members of the family that are not hardy, the shade-loving species of *Clivia*, all from southern

Amaryllidaceae, *Cyrtanthus epiphyticus*

Amaryllidaceae, *Cyrtanthus flanaganii*

Africa, have a rhizome rather than a bulb and are evergreen; they can readily be overwintered indoors. The most desirable species, *C. miniata*, has particularly attractive orange or yellow blooms.

Amaryllidaceae are easily recognizable by the umbellate inflorescence of large flowers (rarely the flowers are solitary), rootstock with few exceptions a bulb, and the leafless and often thick and fleshy flowering stem together with an inferior ovary. *Agapanthus* is broadly similar but differs in the superior ovary and fleshy, rhizome-like rootstock (also in some Amaryllidaceae). *Agapanthus* is conveniently regarded as belonging to the separate single-genus family Agapanthaceae and is treated thus in this book, but some authorities include it in Amaryllidaceae.

Most genera of the family are poisonous to humans and animals alike. Snails and slugs avoid most Amaryllidaceae, and mice and voles will not eat the bulbs. The poisonous principles are a range of alkaloid compounds. An exception is the lily borer caterpillar (*Brithys crini*), which avidly consumes all parts and can kill the plants by boring into the bulbs. Extracts of the bulbs of the African *Boophone* (Greek, ox killer) were used as arrow poison in southern Africa in past times.

Selected Genera of Amaryllidaceae

Amaryllis • Brunsvigia • Clivia • Crinum • Cyrtanthus • Galanthus • Habranthus • Haemanthus • Hessea • Hippeastrum • Lapeidra • Leucojum • Lycoris • Narcissus • Nerine • Pancratium • Scadoxus • Sprekelia • Stenomesson • Strumaria • Sternbergia • Urceolina • Worsleya • Zephyranthes

ANACARDIACEAE

Sumac or Mango Family

About 950 species in 82 genera

RANGE Mainly tropics and subtropics, also Mediterranean and temperate North America

PLANT FORM Evergreen and deciduous trees and shrubs, rarely perennial herbs; leaves mostly in spirals, mostly compound, pinnate or trifoliolate, occasionally simple; often producing toxic resins

FLOWERS Perfect or unisexual, usually small, often many in terminal panicles; calyx usually of 5 sepals, often united basally; corolla usually with 5 free petals, or absent; stamens 5–10; ovary usually superior, of 3–5 united carpels, often 1-celled with 1 ovule or multilocular, styles separate, or 1

FRUIT Usually a drupe, with hard inner layer enclosing a single large seed rich in oils

Anacardiaceae contribute modestly to temperate horticulture, but some North American species of the genus *Rhus*, or sumac, are grown for their dissected to deeply and irregularly cut leaves and brilliant autumn color. The notorious genus *Toxicodendron*, sometimes included in *Rhus*, causes severe allergic skin reactions when touched. Widespread across North America, this shrub or vine is known variously as poison oak or poison ivy. Contact with the leaves or even stems must be carefully avoided. Similar allergic reactions are known in the southern African *Smodingium*. Bark and resin of other genera are also toxic. The only commonly cultivated member of Anacardiaceae in temperate gardens is *Cotinus*, including *C. coccyrigia*, the smoke tree, cultivars of which are grown for their leaf color, dark red or pale yellow-green. The flowers are inconspicuous, but the large, diffuse inflorescences of tiny flowers give rise to the common name. The family also includes the large woody genus *Searsia* (long included in *Rhus*), species of which are widely grown in southern African gardens, especially the willow-like, drought-tolerant *S. lancea*.

Anacardiaceae include *Pistacia*, grown in areas of Mediterranean climate for its fruit, the pistachio nut. Cashew nuts are produced by the tropical American tree *Anacardium occidentale*. The African tree *Sclerocarya*, the marula (or maroela), also yields an edible and delicious nut, and the fleshy fruit is used in a sweet liqueur. The fallen fruits, fermenting on the ground, are eaten by wild animals, especially elephants and monkeys, which become intoxicated as a result. The tropical fruit mango is produced by the genus *Mangifera*. Asian in origin, the mango is now widely cultivated in the tropics across the world.

Dried fruits of *Rhus coriaria* are ground into the spice sumac, a souring agent used for marinades, salad dressings, and a food garnish. The spice is widely used in eastern Mediterranean, Middle Eastern, and Iranian cuisines. The peppercorn tree (*Schinus molle*), native to Peru, is often the only tree that provides shade in dry parts of Australia, California, and southern Africa. The pale red to pink fruits are the source of pink peppercorns (true pepper is obtained from berries of the tropical genus *Piper* in the largely tropical family Piperaceae). Both *S. molle* and *S. terebinthifolius*, the Brazilian peppercorn tree, tend to be invasive in warm temperate and tropical climates.

Anacardiaceae are recognizable by the compound leaves (at least in species grown in temperate climates), five-partite flowers with calyx often united basally and corolla sometimes absent, a superior ovary, and fleshy fruit, a drupe, with a hard inner layer containing a single seed.

Selected Genera of Anacardiaceae
Anacardium • Cotinus • Heeria • Lannea • Mangifera • Pistacia • Rhus • Schinus • Sclerocarya • Searsia • Spondias • Toxicodendron

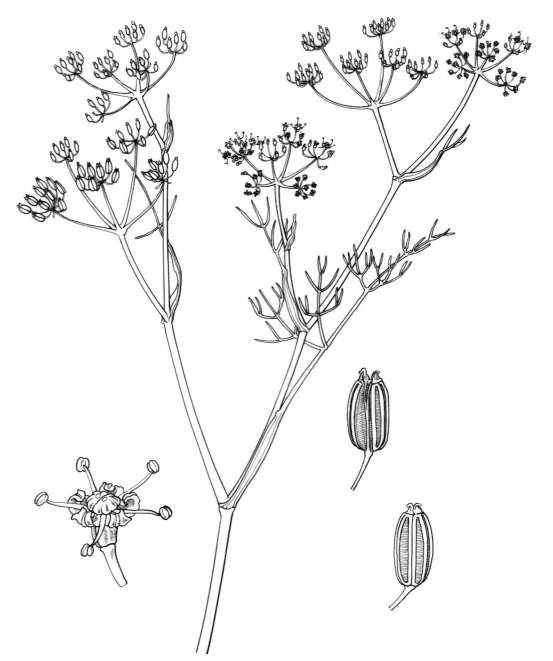

Apiaceae, *Foeniculum vulgare*. As in nearly all Apiaceae, fennel has relatively inconspicuous flowers arranged in umbels. Each flower has five free sepals and petals, five stamens, and an inferior ovary of two carpels with separate styles. Typical of many genera, the petals are keeled on the inner surface with their tips bent inward. The fruits, which are crowned with the persistent calyx, nectar disk, and styles, are ribbed, with oil ducts in the furrows the ribs that contain the aromatic oils that make so many species valuable culinary spices.

APIACEAE

Carrot, Celery, or Parsley Family

About 3700 species in 430 genera

RANGE Cosmopolitan but especially northern hemisphere and tropical mountains, many cold hardy

PLANT FORM Mostly perennial herbs, some evergreen or deciduous shrubs, a few small trees, and annuals, often with ribbed, pithy stems hollow between the nodes, usually aromatic, sometimes poisonous; leaves compound or sometimes simple, then often highly dissected, usually in spirals, often with broad sheathing base

FLOWERS Small, in simple or compound umbels, rarely in head-like clusters, then surrounded by a whorl of free or fused bracts, rarely umbels reduced to 1 flower, perfect or rarely unisexual, radially symmetric or outer flowers of umbels with 1 or more enlarged petals, often white, sometimes yellow or pink, occasionally other colors; calyx of 5 small, usually scale-like or vestigial sepals reduced to thickened rim or sometimes absent; corolla of 5 (usually) free petals with short crest, then curved inward at the tips, usually rapidly deciduous; stamens 5, inserted at base of petals, stamens and petals shed before stigmas mature; ovary inferior, 2-celled with 1 ovule in each cell, with 2 separate styles

FRUIT A dry, hard and sometimes woody structure (mericarp), splitting into two separate units, longitudinally ridged or winged with resin ducts between the ridges, sometimes armed with hooks, bristles, or tubercles

Apiaceae (historical name Umbelliferae) include a fair number of ornamental garden plants, notably *Ammi maius* (Queen Anne's lace) and species of *Angelica*, *Pimpinella*, and *Selinum*. A bronze-leaved strain of fennel (*Foeniculum*) and a dark purple leaved cultivar of chervil (*Anthriscus*) have considerable ornamental merit. Masterwort (*Astrantia*) is especially attractive with large colored bracts surrounding an almost head-like umbel of tiny flowers. Several *Eryngium* species (sea holly), all of which have sessile flowers crowded in heads surrounded by silver, white, or blue bracts, and sculptural, thistle-like foliage, make unusual, sometimes striking garden subjects. More important Apiaceae are those yielding edible food, most significant of which are the swollen taproots of carrots (*Daucus*) and parsnips (*Pastinaca*), and the fleshy leaves or leaf bases of fennel (*Foeniculum*) and celery (*Apium*). Celeriac is the turnip-like root of a cultivar of *Apium graveolens*.

The family is rich in aromatic oils, gums, and resins and is of considerable economic value as culinary herbs and spices, both the fruits and the dried foliage. Important among these are anise (*Pimpinella anisum*), caraway (*Carum carvi*), coriander (the foliage is called cilantro or dahnia; *Coriandrum sativum*), cumin (*Cuminum cyminum*), dill (*Anethum graveolens*), fennel (*Foeniculum vulgare*), and parsley (*Petroselinum crispum*). The flavoring asafetida is obtained from the

Apiaceae, *Heracleum mantegazzianum*

Apiaceae, *Tordylium officinale*

Apiaceae, *Astrantia major*

Apiaceae, *Pleurospermum candollei*

Apiaceae, *Ferula*

gum from taproots of *Ferula assafoetida* and some other *Ferula* species.

Other Apiaceae are poisonous. Hemlock (poison or spotted hemlock, *Conium maculatum*), not the evergreen coniferous tree *Thuja*, also called hemlock, was famously brewed as a potion to execute the condemned Greek philosopher Socrates. *Oenanthe* (water hemlock or water parsley) is also poisonous to humans, and extracts of some species of the genus are added to streams and pools to kill or stun fish. The active ingredients of poison hemlock include coniine, which causes paralysis of the respiratory muscles. Contact with some Apiaceae causes dermatitis, especially *Deverra*, *Heracleum*, and the South African *Notobubon* (or *Peucedanum*) *galbanum*, the last causing a particularly severe reaction comparable to that of poison oak (*Toxicodendron*) in North America. *Centella asiatica* (Indian pennywort) is an important commercial medicinal plant, widely used in wound treatment, especially to prevent postoperative scar tissue formation.

Apiaceae are such a distinctive family there is little confusion about its recognition. Mostly herbaceous, the simple or compound umbels of tiny flowers are unmistakable; the few genera with head-like inflorescences can easily be mistaken for Asteraceae. Individual flowers have free petals, often with the tips curved inward, and two separate styles, which easily separate them from any Asteraceae. Apiaceae are closely allied to Araliaceae, a family that may also have inflorescences of simple or compound umbels. Usually trees or large shrubs, Araliaceae also frequently have fleshy fruits, either drupes or berries, thus unlike any Apiaceae. The two families have at times been united but are currently regarded as separate though immediately related.

Selected Genera of Apiaceae

Alepidea • Ammi • Angelica • Anthriscus • Apium • Astrantia • Berula • Bupleurum • Centella • Conium • Coriandrum • Cryoptotaenia • Cuminum • Cyclospermum • Daucus • Eryngium • Ferula • Foeniculum • Hacquetia • Heracleum • Meum • Oenanthe • Orlaya • Pastinaca • Petroselinum • Peucedanum • Pimpinella • Selinum • Smyrnium • Tordylium • Torilis

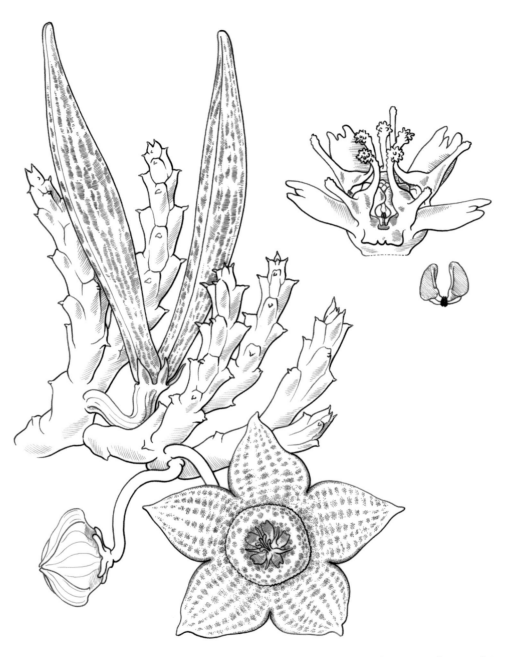

Apocynaceae, *Orbea variegata* (synonym *Ceropegia mixta*). Carrion flower is one of a group of stem succulents in the family in which the opposite leaves are reduced to vestiges. The five sepals are more or less free, but the leathery petals are joined in a short tube that encloses the superior ovary and the stamens. These are united into a complex structure termed a gynostegium, in which the pollen grains of adjacent anthers are amalgamated into small waxy parcels called pollinia. The two separate carpels of the ovary mature into a pair of characteristic horn-like fruits. The large, rather evil-looking flowers, smelling of rotting fish, attract carrion flies as their pollinators. Unlike most Apocynaceae, *Orbea* and *Ceropegia* lack milky latex.

APOCYNACEAE

Dogbane or Milkweed Family

About 4700 species in 345 genera

RANGE Largely tropical and warm temperate but some Mediterranean and a few
cold temperate

PLANT FORM Trees, shrubs, vines, and perennial herbs or geophytes, a few annuals; some
stem succulents with vestigial leaves or without leaves; leaves (when present) simple, opposite or in close whorls, occasionally in spirals, sometimes succulent, or lacking, sometimes with
small stipules, usually hairless and glossy; usually with milky or colorless latex

FLOWERS Perfect, radially symmetric, borne in racemes, cymes, bracteate umbels, or solitary;
calyx of 5 free or basally united sepals; corolla of 5 free petals or united below in a cup or tube
often salver-shaped, with 4 or 5 lobes, sometimes with a petal-like projection from the base,
the corona, enclosing the stamens (as in subfamily Asclepiadoideae), lobes either twisted like
an umbrella in bud or abutting; stamens 5, inserted on corolla tube, free or anthers or entire
stamens adhering to style or style head, pollen sometimes forming a single mass (a pollinium)
in each anther lobe; ovary usually superior, of 2 carpels basally united or free but with 1 style
and large, variously elaborate stigma

FRUIT Mostly elongate units (follicles) borne in pairs, containing many seeds with prominent
tuft of fine, silky hairs, or fleshy

A comparatively large family, Apocynaceae are subdivided into several subfamilies, the most important from a horticultural point of view being Asclepiadoideae,
often in the past treated as a separate family, Asclepiadaceae, the true milkweeds.
Several species of *Asclepias* are hardy in temperate gardens, and many more
species of several genera, especially the stem succulents, are grown in areas
of Mediterranean and warm desert climates. Numerous genera and species are
grown in specialist collections indoors or in greenhouses. Relative to its size,
Apocynaceae contribute only modestly to temperate gardens. *Asclepias curassavica* has striking orange and yellow flowers and is attractive to butterflies. The
North American genus *Amsonia* is hardy and occasionally grown in gardens. *Vinca
major* and *V. minor* (the periwinkles) are widely used as ground covers in temperate gardens although both tend to be invasive and may be hard to eradicate. Species
of the spiny southern African shrub *Carissa* are cultivated for their scented white
flowers and delicious dark red fruit. Their thorny branches also make them a useful
hedge plant. The vine *Trachelospermum jasminoides* (star jasmine) is hardy and
grown in gardens that receive hard freezes but is more at home in warmer climates.
Its starry, jasmine-like flowers are wonderfully scented. *Apocynum*, with that
curious common name dogbane, is a milkweed look-alike. Species contain a cardiac
toxin and the generic name, which is of Greek origin (dog death), refers to its pre-
sumed toxicity to dogs and, we assume, other animals.

The Madagascan *Catharanthus roseus*, often called Madagascar periwinkle, is a small shrub with attractive pink or purple flowers, hardy only in tropical and Mediterranean gardens but grown as an annual in temperate gardens. Ironically, *C. roseus* is endangered in the wild. As the common name suggests, its flowers resemble those of *Vinca*, to which it is related. The frangipani (*Plumeria*) is widely grown in tropical and Mediterranean gardens; its scented flowers are used in garlands in Polynesia, including Hawaii. Another shrub, *Nerium oleander*, commonly called oleander, is used in street plantings and gardens in warm temperate parts of the world although it is native only to southern Europe and the Middle East. Ingestion of a small quantity of its leaves will kill an ox, and use of its twigs as kebab skewers is reputed to have had fatal consequences for some of Alexander the Great's soldiers. The extremely bitter sap deters attempts to eat any parts of the plant. Species of the genus *Adenium* are widely cultivated in the Far East as an ornamental. One species of the genus *Cryptostegia*, originally from Madagascar, is a common weed in tropical parts of southern Africa and India.

Many Apocynaceae are sources of drugs and poisons. *Strophanthus* species yield strophanthine, used in treatment of heart disease. Reserpine, an antihypertensive and antipsychotic drug, is obtained from species of the tropical genus *Rauvolfia*. An appetite suppressant is obtained from the southern African genus *Hoodia*, and the sap of *Acokanthera* was used in the past as an arrow poison by African hunters. *Catharanthus roseus* is also poisonous, but extracts have been used medicinally for centuries against various diseases, notably malaria. The compounds vinblastine and vincristine extracted from the plant are used today for the treatment of leukemia and Hodgkin's lymphoma.

Selected Genera of Apocynaceae

Acokanthera • Adenium • Allamanda • Amsonia • Apocynum • Asclepias • Brachystelma • Caralluma • Carissa • Ceropegia • Cryptostegia • Cynanchum • Dischidia • Duvalia • Hoodia • Hoya • Huernia • Gomphocarpus • Mandevilla • Nerium • Orbea • Oxypetalum • Plumeria • Rauvolfia • Sarcostemma • Stapelia • Strophanthus • Thevetia • Trachelospermum • Vinca

Apocynaceae, *Strophanthus speciosus*

Apocynaceae, *Asclepias curassavica*

Apocynaceae, *Vinca herbacea*

Apocynaceae, *Adenium obesum*

Araceae, *Zantedeschia rehmannii*. The relatively broad, spear-shaped leaves with a conspicuous petiole are characteristic of many members of the family. The large petaloid structure, or spathe, on the flowering stem is a modified, colorful bract subtending a finger-like column, the spadix, bearing the numerous, tiny, unisexual flowers, each reduced either to a solitary sessile anther or an ovary containing just one ovule. In *Zantedeschia* the female flowers are restricted to the lower part of the spadix, with the male flowers above them. Pollen grains are exuded from tiny pores in the anthers.

ARACEAE

Arum Family

About 6450 species in 119 genera

RANGE Nearly cosmopolitan but largely tropical and subtropical, also temperate Eurasia

PLANT FORM Evergreen scrambling shrubs, vines, or perennial herbs, or seasonal geo-phytes with rhizomes, tubers, or corms (sometimes very large), some rooted or floating aquatics; leaves simple, in spirals or two-ranked, parallel- or net-veined, often with distinct petiole clasping at base, blades sometimes lobed, dissected, or with holes, petiole often with spongy pith

FLOWERS Many, small, without bracts, arranged on an elongate or short axis (the spadix), this usually subtended and partly enclosed by a large, colored leafy organ (the spathe); bisexual or unisexual, perianth not differentiated into calyx and corolla, thus with 4–6 free tepals in 2 whorls, or tepals lacking; stamens mostly 4, 6, or 8, anthers opening by pores, narrow slits, or longitudinal slits; ovary superior, of (2 or) 3 (to many) united carpels, styles short or lacking, locules with 1 to many ovules; inflorescences sometimes produced before leaves, and male and female flowers separated on spadix

FRUIT Usually a berry, rarely dry, sometimes the entire spadix forming a fleshy, compound fruit

As with most largely tropical families, Araceae offer few plants for temperate gardens. Many more are grown in greenhouses, as potted plants for the home, or simply as annuals, replaced each year. The so-called flowers are actually inflores-cences. The large colored organ, the spathe, subtends and often surrounds a stalk, the spadix, bearing many small flowers. Often the flowers are unisexual with males and females separated on different parts of the spadix. Several seasonal geophytes from Eurasia are hardy, some only in western and southern North America and the Mediterranean, where they are locally and somewhat indiscriminately called jack-in-the-pulpit. Among these, pride of place goes to *Arisaema*, a genus of more than 150 species. Several from eastern Asia and North America are prized in gardens for their striking and rather ominous looking inflorescences, the spadix sometimes with tip blunt and rounded or, instead, elongate and trailing. *Arum italicum* and *A. maculatum* are widely grown for their attractive inflorescences and long-lasting red fruit, ripening in the autumn when the foliage has died back. The eastern North American *Orontium aquaticum* (golden club) is a useful plant for ponds and water-logged areas. The inflorescence has a vestigial spathe, and display is provided by the large white spadix tipped with golden yellow.

The southern African genus *Zantedeschia*, better known as the arum or calla lily, includes *Z. aethiopica* with large white spathes and which is widely grown and even weedy in places; several other species of *Zantedeschia* or their hybrids, with yellow

Araceae, *Dracunculus vulgaris*

Araceae, *Lysichiton americanum*

Araceae, *Arum italicum*

Araceae, *Arisaema jacquemontii*

or pink to red spathes, are available in the horticultural trade. The skunk cabbages, so named for the foul odors produced when the flowers are mature, are also hardy and merit a place in gardens, especially suited to poorly drained sites. *Symplocarpus foetidus*, the true skunk cabbage, native across North America and northern Asia, has purple to black or mottled spathes of dramatic appearance produced before the leaves emerge. When flowers are mature and ready for pollination the spadix generates heat and scent, often extremely unpleasant, to attract pollinators, frequently flies or beetles. *Lysichiton*, false skunk cabbage, has two species: the western North American *L. americanum* with large, bright yellow spathes, and *L. camtschatcensis* from eastern Asia with pure white spathes. Both are hardy and useful additions to gardens.

Araceae suitable for indoor culture include the large-leaved *Dieffenbachia*, *Monstera*, and several species of *Philodendron*, all grown mainly for their foliage, and *Anthurium*, which produces attractive and long-lasting inflorescences with red, pink, or purple spathes. Leaves of many Araceae contain quantities of oxalate crystals, rendering them inedible and somewhat toxic. Pond weeds, often called duckweed, *Lemna*, *Spirodela*, and *Wolffia* are floating plants with reduced plant bodies and tiny inflorescences. They can cover the surface of pools and slow-moving bodies of water.

Araceae are of great economic importance as food plants in the tropics, the starchy roots or tubers providing a food staple for millions of people. *Colocasia*, the taro plant, is grown throughout Indonesia and Polynesia and has spread from there to other tropical lands. Now with many hundreds of cultivars, taro has been cultivated for more than 10,000 years.

Selected Genera of Araceae
Alocasia • Amorphophallus • Anthurium • Arisaema • Arisarum • Arum • Caladium • Colocasia • Dracunculus • Dieffenbachia • Lemna • Lysichiton • Monstera • Orontium • Philodendron • Pistia • Syngonium • Symplocarpus • Zantedeschia

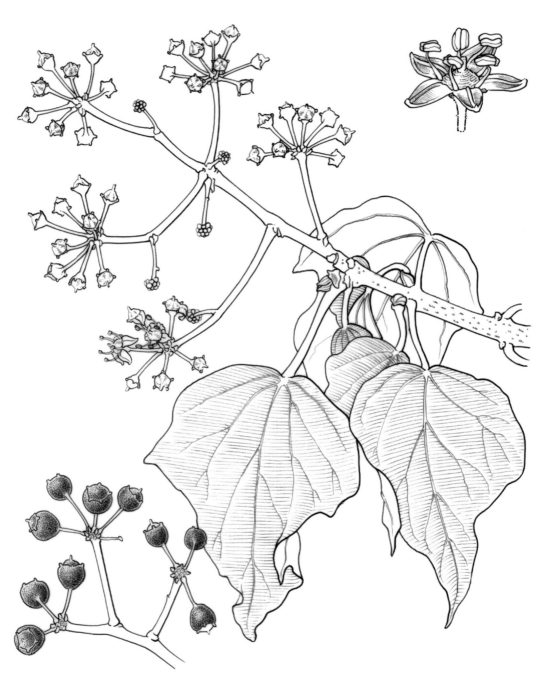

Araliaceae, *Hedera helix*. As in Apiaceae, flowers of Araliaceae are often borne in compound umbels. Individual flowers have five free sepals and petals, five stamens, and an inferior ovary of two single-seeded carpels. In *Hedera* there is a single style.

ARALIACEAE

Ginseng or Aralia Family

About 1900 species in 40 genera

RANGE Nearly cosmopolitan but largely tropical

PLANT FORM Small to large, mostly evergreen trees and shrubs, some with pachycaul habit, some epiphytes, vines, and perennial herbs; leaves compound, pinnate or digitate, or simple and palmately lobed, in spirals or crowded at the tops of branches, sometimes with prickles on the leaf stalks and stems, usually with stipules

FLOWERS In simple or compound umbels, racemes, spikelets, or heads, usually small, greenish or white; perfect or unisexual; perianth with calyx of 4 or 5 small tooth-like sepals (or obscure or absent) and mostly 4 or 5 free petals keeled on upper surface; stamens usually as many as petals and alternating with them, or twice as many, or numerous, anthers with longitudinal slits; ovary almost always inferior, of 2–5 (to many) united carpels and as many locules and styles, or styles sometimes partly or fully united

FRUIT Most often a berry or drupe, rarely dry, not splitting into segments (as in related Apiaceae)

Closely related to the carrot family (Apiaceae), Araliaceae share with that family the distinctive umbellate inflorescences, at least in most species. Araliaceae differ, however, in their mostly woody habit and often compound or sometimes palmately lobed or palmately veined leaves (some Apiaceae have compound leaves), and in lacking resin ducts in the fruits. The overwhelming number of genera and species are tropical and have no place in temperate gardens. One striking exception is the genus *Hedera* (for example, *H. helix*, ivy), commonly grown as a ground cover or vine. The ivy seen in most gardens and roadside plantings is in the juvenile state, but as plants mature they produce upright branches bearing characteristic *Aralia*-like umbels of small flowers followed by fleshy, berry-like fruits. Plants are dispersed by birds, which consume the fleshy fruits. Another genus in cultivation is the shrub *Fatsia*, which has large, glossy, palmately lobed leaves. The attractive and hardy hybrid between ivy and fatsia is called ×*Fatshedera*.

Aralia spinosa, native to eastern North America, is hardy and an attractive small tree with relatively large (for the family) pink flowers in massive compound umbels. Several species of *Aralia* are perennial herbs and truly hardy. Among these is the California aralia, sometimes called elk clover, *A. californica*, a fine garden subject. A strain with chartreuse-yellow foliage is especially attractive. Several Chinese species of Araliaceae are seen occasionally in gardens, notably *Tetrapanax papyrifera*, the Chinese pith paper tree; it has huge, lobed leaves crowded at the tops of unbranched stems. Several more species from China and some from New Zealand and South America are hardy or half-hardy and have considerable horticultural merit. Among these are *Metapanax davidii*, *M. delavayi*, *Pseudopanax laetivirens*, and *Schefflera taiwaniana*. Araliaceae also provide a range of attractive houseplants. Two particularly commonly grown plants, both from New Caledonia, are *Plerandra elegantissima* and *P. veitchii*, better known by their older names *A. elegantissima* and *A. veitchii*.

The genus *Panax* includes perennial species prized as the medicinal herb ginseng, produced from roots harvested from the eastern Asian *P. ginseng* and *P. pseudoginseng* and the North American *P. quinquefolius*. Ginseng is used as a tonic, sometimes an aphrodisiac, and a treatment for cancer, colds, diabetes, and sexual dysfunction among other maladies. Pith paper is made from the soft, white pith of *Tetrapanax*, actually native to China and Taiwan but widely cultivated elsewhere. Bark or roots of some species are used locally as medicines, and young leaves of others as a vegetable, especially those of *Eleutherococcus*.

Cultivated species of Araliaceae are recognizable by the compound or palmately lobed leaves, large umbellate (less often spike-like) inflorescences of small, usually inconspicuous flowers with free sepals (often so small as to be obscure) and petals, stamens usually as many or twice as many as petals, an inferior ovary, and fleshy fruit. There are usually more than two styles, or the styles are united completely below (related Apiaceae have two free styles, and dry fruit splitting into two sections).

Selected Genera of Araliaceae
Acanthopanax • *Aralia* • *Cussonia* • *Eleutherococcus* • *Fatsia* • *Hedera* • *Metapanax* • *Oplopanax* • *Panax* • *Plerandra* • *Polyscias* • *Pseudopanax* • *Schefflera* • *Tetrapanax*

ARECACEAE

Palm Family

About 2550 species in 184 genera

RANGE Tropical and subtropical, a few in Mediterranean climates

PLANT FORM Evergreen trees, shrubs, and vines; leaves in spirals or two-ranked, usually in terminal rosettes, simple but sometimes appearing compound, pinnately divided or palmate (fan palms), rarely entire, with parallel venation

FLOWERS Bisexual or unisexual, often small but usually in large, axillary panicles, 3-merous, radially symmetric, often creamy yellow; perianth of 3 free outer and 3 free inner tepals; stamens often 6 in 2 whorls or many, anthers often with lateral longitudinal slits; ovary superior, mostly of 3 united carpels with as many cells, each with 1 ovule, with free styles or 1 style (or stigma sessile)

FRUIT Usually a fleshy or fibrous drupe (for example, dates)

Arecaceae, *Trachycarpus fortunei*

Of limited importance to temperate horticulture, just a handful of palms (Arecaceae, historical name Palmae) are hardy in warm temperate climates, not in cold climates. Nevertheless, their stately and sculptural forms remain a temptation for many gardeners. The western Mediterranean *Chamaerops humilis*, one of the smaller palms, is readily grown in coastal temperate climates. The native southwestern North American *Washingtonia* is used as a street tree; *W. filifera* is the palm of Palm Springs in interior California. The genus *Trachycarpus*, windmill palms, and especially *T. fortunei*, native to eastern China, is often grown in Mediterranean and warmer temperate climates as sculptural garden subjects and in street plantings. Several small palm species, especially of the genus *Rhapis*, are grown indoors as pot plants and of course in greenhouses. Their natural habitat as understory plants in rain forests renders them ideal for situations of low light intensity.

Palms are closely associated with humans in the tropics, providing food, wood, and fiber for furniture, cloth, and basketry. The coconut (*Cocos nucifera*) provides food and is a staple in tropical islands and coastal areas and also a valuable export crop. Native to western Africa, the oil palm (*Elaeis guineensis*) is a rich source of a saturated oil and is now widely cultivated in the Asian tropics and Madagascar to the detriment of native forest and wildlife. Palm oil is used in soap, candles, cosmetics, foods including margarine, and even as fuel for vehicles. The date palm (*Phoenix dactylifera*), probably native to Arabia and northern Africa, is a cultigen, not known in the wild. It has been cultivated in the ancient Middle East for at least 6000 years and is a source of fresh and dry fruit, sugar, and fiber for matting. The species is also used as an ornamental in areas of Mediterranean tropical climate, sometimes as a street tree. The African Senegal palm (*P. reclinata*) and Macaronesian Canary palm (*P. canariensis*) are also cultivated in the tropics and warm temperate zone, especially as street trees. The sap is tapped in Africa to make a palm wine. Heart of palm is obtained from the soft, fleshy buds of several palm species. Sap from *Corypha*, *Nypa*, and other palms is evaporated to produce sugar or fermented and distilled to make arrack.

Selected Genera of Arecaceae
Areca • Bactris • Borassus • Caryota • Chamaerops • Cocos • Corypha • Elaeis • Hyphaene • Livistona • Nypa • Phoenix • Raphia • Rhapis • Sabal • Serenoa • Trachycarpus • Washingtonia

ARISTOLOCHIACEAE

Birthwort or Wild Ginger Family

About 660 species in 8 genera

RANGE Tropical and temperate, especially the Americas

PLANT FORM Shrubs, woody vines, and perennial herbs, many with creeping rhizomes; leaves in spirals, simple, often heart- or kidney-shaped, often with palmate venation, sometimes aromatic when crushed, without stipules

FLOWERS Bisexual, in racemes, cymes, axillary clusters or solitary, sometimes zygomorphic, often dull colored and speckled; calyx of 3 often petaloid lobes partly to largely united in S-shaped tube; corolla of 3 petals, often small, sometimes lacking; stamens often 6, or many, occasionally 1–5, free or united with style; ovary superior or inferior, of (4 to) 6 united carpels with as many cells, each with many ovules and 1 style (or carpels almost free with separate styles in *Saruma*)

FRUIT Usually dry, either capsules or when from separate ovaries, follicles (*Saruma*), or not spitting

Aristolochiaceae, *Saruma henryi*

Aristolochiaceae, *Asarum marmoratus*

Of minor importance to horticulture, Aristolochiaceae are perhaps best known for the vine *Aristolochia* (for example, Dutchman's-pipe), several species of which have distinctive flowers with an S-shaped tube formed from a petaloid calyx and lacking a true corolla. The Eurasian *A. clematitis* (birthwort) was so named for the shape of the tube, said to resemble a fetus in the correct position prior to birth (thus thought to aid birth but in fact causing abortion).

Species of *Asarum* (wild ginger) are prostrate herbs, usually spreading on short to long rhizomes. The flowers, mostly unusually colored brown or purple and cream, are held beneath the leaves and thus usually overlooked although quite large. The leaves are often aromatic when crushed, some smelling of ginger, hence the common name. Several species are cold hardy and make interesting ground covers, some with mottled leaves. Not often seen in gardens, the Chinese *Saruma henryi* (upright wild ginger) is

a perennial with conspicuous flowers, with a bright yellow corolla of three large, free petals. Unusual in Aristolochiaceae, *S. henryi* has nearly free carpels with the stamens not appressed to the style.

Aristolochiaceae are recognizable by the kidney- to heart-shaped leaves often on long petioles, three-merous flowers, often with the corolla reduced or absent, thus a perianth of just one whorl of three united sepals sometimes in an S-shaped tube. The stamens are often appressed to the style, although free in *Saruma*, which is also unusual in having a well-developed corolla, a calyx of small sepals, and carpels free terminally and with separate styles. Species contain alkaloids and are poisonous, some are potent carcinogens, and others are used medicinally as an emetic or abortifacient.

Selected Genera of Aristolochiaceae
Aristolochia • Asarum • Holostylis • Isotrema • Saruma

ASPARAGACEAE

Asparagus or Agave Family

About 1650 species in 85 genera

RANGE Cosmopolitan, especially semiarid parts of the world

PLANT FORM Small trees often with pachycaul habit, shrubs, a few vines, and many perennials; rootstock often a rhizome or a bulb or corm, some with fleshy, swollen roots; leaves in spirals or basal rosettes, simple, entire, sometimes large and strap-like, sometimes with spiny tips, or leafless with flattened or needle-like stems resembling leaves and the true leaves reduced to colorless scales, without stipules

FLOWERS Usually bisexual, sometimes unisexual, in racemes, spikes, or axillary clusters; perianth with the outer whorl petal-like and not differentiated into a calyx, thus consisting of 3 outer and 3 inner tepals, free or united and tubular below, variously colored but often white or yellow, occasionally red, rarely blue; stamens 6, as many as tepals, filaments sometimes united in a tube around style; ovary superior or inferior, consisting of 3 united carpels, usually 3-celled, with 1–12 ovules per locule; style usually 1

FRUIT A dry capsule or fleshy berry (especially *Asparagus*); seeds in dry capsules usually flattened and with black seed coat, sometimes with a fleshy white appendage (elaiosome)

Circumscription of Asparagaceae has changed since the 1980s and expanded to include families regarded in the past as Agavaceae, Anthericaceae, Convallariaceae, Dracaenaceae, Hostaceae, Ruscaceae, Themidaceae, several genera of Liliaceae, and more that need not be mentioned here. Many genera are largely tropical or from warm deserts, but several are cultivated in cool temperate or Mediterranean climates. These include *Agave* and *Yucca* as well as species and cultivars of the Asian genus *Hosta*, all of which are cold hardy. The tree-like *Y. brevifolia* (Joshua tree) has an unusual growth form for the family, with a tree-like trunk and several thick branches bearing terminal rosettes of sword-like leaves. The tuberose (*A. tuberosa* or *Polianthes tuberosa*), grown as a cut flower, has long-lasting, sweetly scented blooms. *Convallaria* (lily-of-the-valley) is a much-loved spring-blooming perennial with delightfully scented flowers. Species of the genus *Dracaena*, like some *Yucca* species, have woody stems, and some species, most notably the dragon tree (*D. draco*) from the Canary Islands, are tree-like. This and a few other *Dracaena* species produce a red resin called dragon's blood, giving rise to their common name. Dragon's blood was employed in antiquity to staunch bleeding and is today used in stains and varnishes.

The liquor tequila is distilled from the fermented sap of some Mexican *Agave* species. Several species of Asparagaceae are edible, most notably *Asparagus officinalis*, which is widely cultivated; the spring shoots are harvested as a vegetable. Leaves of some *Hosta* species are eaten in eastern Asia. Species of

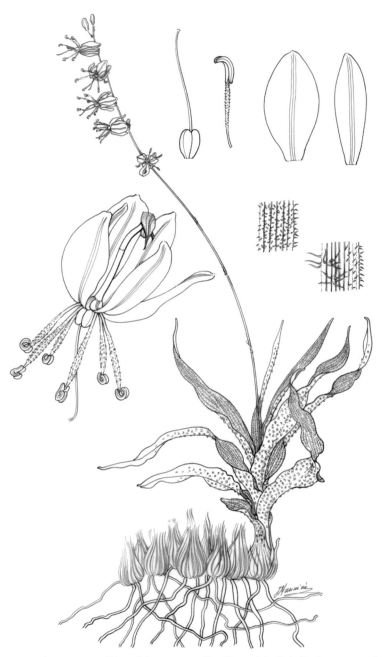

Asparagaceae, *Chlorophytum asperum*. The rhizome is typical of herbaceous members of the family. Also typical are the spirally inserted, strap-like leaves with parallel venation, and the raceme of regular flowers. Each flower of *C. asperum* has six white, petal-like corolla segments, six free stamens, and a three-lobed superior ovary with a single style. Although the flowers in many genera are long lived, those of *Chlorophytum* are fleeting, lasting less than a day. The barbed filaments are characteristic of many species in the genus but not of the family.

Asparagaceae, *Anthericum liliago*

Asparagaceae, *Brodiaea howelliae*

Asparagaceae, *Camassia quamash*

several other genera are grown in gardens, notably the North American *Camassia* (camas), *Polygonatum* (Solomon's seal), and the western North American *Brodiaea* and *Dichelostemma*.

The North American species included in the past in *Disporum* are now the genus *Prosartes;* as circumscribed today, *Disporum* is exclusively Asian and is included in the family Colchicaceae, only distantly related to Asparagaceae. Hyacinthaceae, with some 800 species, is sometimes included in Asparagaceae, but we deal with that horticulturally important family separately in the A–Z.

Selected Genera of Asparagaceae

Agave • Aloiampelos • Anthericum • Arthropodium • Asparagus • Aspidistra • Beschorneria • Brodiaea • Camassia • Chlorogalum • Chlorophytum • Clintonia • Convallaria • Danae • Dichelostemma • Dracaena • Hosta • Liriope • Maianthemum • Nolina • Ophiopogon • Polygonatum • Prosartes • Ruscus • Sansevieria • Semele • Yucca

ASTELIACEAE The agave-like *Astelia* (known by the evocative common name pineapple grass) from Australasia is assigned to the small family Asteliaceae, the leaves of which often have distinctive silvery or white hairs. The seeds, like those of Asparagaceae, have a black coat.

ASPHODELACEAE

Aloe Family

About 1050 species in 39 genera

RANGE Mainly African, also Europe to central Asia

PLANT FORM Mainly shrubs and perennials, many rosette-forming succulents, a few tree-like, some tuberous geophytes; leaves simple, strap-like or succulent, often with prickles and variously shaped; usually with sticky sap with medicinal properties

FLOWERS Bisexual, mostly radially symmetric, in spikes, racemes, or panicles; calyx and corolla not differentiated, both petal-like, thus tepals 6 in 2 whorls, free or united and often tubular; stamens 6 in 2 whorls, often unequal in length; ovary superior, of 3 united carpels, 3-celled, with 1 style, with few to many ovules

FRUIT Dry capsules, with angular or flattened seeds with black seed coat, covered with a thin membranous outer skin, giving a dull surface with pale flecks of minute oxalate crystals, winged in *Eremurus*

Probably best known among Asphodelaceae is the large genus *Aloe*, with more or less succulent leaves and thus mostly not frost hardy, but several species are grown in gardens with Mediterranean to frost-free climates, and many more are cultivated under glass in specialist collections. Most *Aloe* species are stemless, but the genus also includes some unbranched, tree-like species. The profusely branched, tree-like aloes are now considered to belong to a second genus *Aloidendron*, while the trailing, vine-like species have been segregated as the genus *Aloiampelos*. Smaller, stemless leaf succulents include *Astroloba*, *Gasteria*, with reddish, urn-shaped, bird-pollinated flowers, and *Haworthia* and the closely related *Haworthiopsis* and *Tulista*. Species of the African genus *Kniphofia* (red hot poker, which has strap-like leaves, usually not succulent) are increasingly seen in temperate gardens; the hardy selections and cultivars are derived from species of high elevations in the southern African Drakensberg. The Asian *Eremurus* species (foxtail lilies) are fine garden subjects, as is the white-flowered Eurasian *Asphodelus*. African *Bulbine* and *Bulbinella* are also grown in gardens in Mediterranean or warm climates.

Extracts of *Aloe* sap, especially from *A. vera* and *A. ferox*, are used medicinally and in the cosmetics industry. *Asphodelus* provides dyes and gums.

Among the lily-like families with a petaloid perianth of six tepals, six stamens, and a superior ovary, Asphodelaceae are recognizable by the usually succulent leaves (not *Kniphofia*), sometimes with prickles, and slimy, often yellow, bitter-tasting sap; flowers usually arranged in racemes or spikes; and the dull black seeds covered by a thin membrane containing flecks of calcium oxalate crystals.

Asphodelaceae, *Aloiampelos tenuior.* The attractive, yellow or orange flowers of the genus are adapted to pollination by nectar-feeding sunbirds. Like the allied Asparagaceae, the sepal and petal whorls are both petaloid, hence called a perianth, often tubular and partly enclosing the style and six stamens. The succulent leaves are armed along the margins with soft prickles. The outer green part of the leaves contains a distinctive, bitter yellow exudate, but the colorless central part is used cosmetically.

Asphodelaceae, *Kniphofia*

Asphodelaceae, *Asphodelus luteus*

Asphodelaceae, *Asphodelus ramosus*

Selected Genera of Asphodelaceae

Aloe • Aloiampelos • Aloidendron • Asphodeline • Asphodelus • Astroloba • Bulbine • Bulbinella • Eremurus • Gasteria • Haworthia • Haworthiopsis • Kumara• Tulista • Trachyandra

The Australian genera *Kingia* and *Xanthorrhoea*, sometimes included in Asphodelaceae, are more often segregated as **Dasypogonaceae** and **Xanthorrhoeaceae**, respectively. Commonly known as black boys, the plants have grass-like foliage, which is unusual for Asphodelaceae.

HEMEROCALLIDACEAE Also sometimes included in Asphodelaceae, Hemerocallidaceae (daylily family) include *Hemerocallis*, the hardy daylily, with 15 species from eastern Asia; *Phormium* from Australasia;

Dianella from Asia and Australasia; and a few more. The most well known daylily is the orange-flowered *H. fulva*, the parent of thousands of named cultivars, mostly color variants but some with doubled flowers. *Hemerocallis lilioasphodelus* (*H. flava*) has smaller yellow flowers with a pleasant scent. *Phormium* (New Zealand flax) is grown mainly for its long-lived and elegant sword-like foliage, and thrives best in areas of warm temperate or Mediterranean climate, in North America mainly in California and Florida. Farther north, plants may not survive more than one or two mild winters unless temperatures remain above 20 degrees Fahrenheit (–7 degrees Celsius).

ASTERACEAE

Daisy Family

About 25,000 species in 1570 genera

RANGE Cosmopolitan, especially arid parts of the world

PLANT FORM Shrubs, small or occasionally large trees, vines, perennial herbs, and annuals; leaves mostly simple, sometimes dissected to more or less compound, often toothed, usually lacking stipules; often aromatic and variously hairy to cobwebby

FLOWERS In bracteate heads of several to many flowers (hence the alternate family name Compositae, indicating a composite flower), these apparent "flowers" are flower heads surrounded by discrete or variously fused bracts in one or more whorls forming the involucre; the actual flowers of the head comprise small individual flowers in the center, the disk florets, often with strap-shaped flowers radiating from the edge of the disk, the ray florets (ray florets may be absent in some genera and species); disk florets perfect or unisexual; calyx represented by scales, hairs, a raised rim above the ovary, or absent; petals united into a funnel-shaped tube below, usually 5-lobed, variously colored but often yellow; ray florets invariably female; calyx represented by hairs, scales, or absent; corolla lobes largely united into a strap-like limb, variously colored but often yellow; stamens (when present) as many as petals; ovary inferior, consisting of 1 carpel with 1 ovule, with style usually divided into 2 lobes (structure of the style important for classification of the major groupings, subfamilies and tribes)

FRUIT Usually dry and with various appendages (hairs, wings, hooks) to aid dispersal, occasionally fleshy as in some shrubby *Osteospermum* species (in the past, the genus *Chrysanthemoides*), one species of which is an invasive weed, called boneseed, in parts of Australia

The second largest family of flowering plants after Orchidaceae, Asteraceae (historical name Compositae) include so many genera of horticultural importance that we can mention only a few. *Aster* in the broad sense is now divided into several genera, most blooming in the late summer and autumn. The genus *Chrysanthemum* includes the widely grown plants loosely called chrysanthemums. The autumn-flowering chrysanthemum of horticulture and the florist trade is a complex hybrid called *C. ×morifolium*, bred in China more than 1500 years ago, but species have been cultivated there for more than 3000 years. The annual called *C. coronaria*, from the Mediterranean, sometimes cultivated but more often encountered as a weedy annual, is now in the genus *Glebionis*. Annuals of garden importance include the so-called Namaqualand daisies from southern Africa in the genera *Dimorphotheca* and *Osteospermum* (for example, *O. calcicola*, limestone window-seed). The South African *Arctotheca* (called Cape weed) is weedy in western North America. Another South African native, the Transvaal or Barberton daisy, *Gerbera*, is important as a cut flower, with hundreds of cultivars in existence.

Asteraceae, *Osteospermum calcicola*. A typical radiate daisy, the outer ring of florets in the head have petals united in a narrow, spreading limb or ray, whereas the central or disk florets are radially symmetric and funnel-shaped. In *Osteospermum* the radiate flowers are female lacking stamens, and fertile, and the disk florets are functionally male, with five stamens. As in all daisies, the stamens are joined along their edges into a column and shed their pollen grains directly onto the elongating style, which pushes the grains up out of the staminal column like a piston.

Asteraceae, *Gazania rigida*

Asteraceae, *Berkheya glabrata*

Asteraceae, *Ligularia fischeri*

Asteraceae, *Tragopogon coloratus*

Of low-growing habit, *Gazania* species, also southern African, and their hybrids are widely grown in warm temperate gardens and thrive in dry, exposed habitats.

The family includes the large herbs loosely called thistles, assigned to several genera, including *Cirsium*, *Cynara* (artichoke and cardoon), and *Echinops* (globe thistle), all of which lack ray florets. The large genus *Senecio* (the old common name groundsel seems redundant today) includes tree-like plants, shrubs, perennials, and many annuals. *Zinnia* and close allies are attractive perennials and annuals bred for larger and more colorful flowers than the wild plants.

The North American genus *Liatris* (gay feather, blazing star) offers attractive hardy perennials with heads of purely disk flowers arranged in attractive spike-like clusters. *Gaillardia* (blanket flower) is another hardy garden favorite of the daisy family. Several native North American species, both annual and perennial, are useful garden plants.

The sunflowers, *Helianthus* species, include several garden ornamentals. *Helianthus annuus* is an important crop grown for oil from the seeds, for example, birdseed. The Jerusalem artichoke is the fleshy underground rhizome of *H. tuberosus*, its inappropriate name a corruption of the Italian for sunflower, *girasole*. Other edible Asteraceae include lettuce, cultivars of *Lactuca sativa*, itself not a wild plant but an ancient cultigen from which the bitter principles have been bred out. The edible artichoke is the flower head of *Cynara cardunculus* harvested before the individual flowers are ready to open. Chicory (*Cichorium intybus*) yields a product of the same name, obtained from the roasted taproot of the plant, used as a substitute or additive to coffee. Plants have attractive blue flowers, but the species is weedy in waste places; the flowers open in the morning and close after midday. Surprisingly, radicchio is a cultivar of the same species, and endive, escarole, and frisé are cultivars of the related *C. endivia*.

Among other useful Asteraceae, the South American *Stevia* yields an organic sweetener used in place of sugar. Some Eurasian species of *Artemisia* (known as wormwood) are used as flavorings, especially in liqueurs, most famously in absinthe. The North American species of *Artemisia*, called sagebrush, most notably *A. tridentata*, are dominant plants over huge areas of the West. The herb tarragon is also a species of *Artemisia*, *A. dracunculus*.

Selected Genera of Asteraceae

Achillea • Arctotis • Arnica • Artemisia • Aster • Bellis • Chrysanthemum • Cichorium • Cirsium • Coreopsis • Cosmos • Cynara • Dahlia • Dimorphotheca • Doronicum • Echinops • Eupatorium • Gaillardia • Gazania • Glebionis • Helianthus • Lactuca • Liatris • Ligularia • Osteospermum • Othonna • Petasites • Ratibida • Sanvitalia • Senecio • Stevia • Zinnia

BEGONIACEAE

Begonia Family

About 1800 species in 2 genera

RANGE Mostly tropics and subtropics, a few warm temperate

PLANT FORM Succulent shrubs, epiphytes, and perennial herbs, some with underground tubers; leaves in spirals or two-ranked, simple and often asymmetric or palmately lobed or compound, with stipules, these often large

FLOWERS Usually unisexual, borne in axillary cymes, often irregular, the female flowers larger than the male; perianth of 2 petaloid whorls, usually free, males with 2 unlike sets of 2 petals, females with a single set of 5 petals; stamens 4 to many, arranged on one side of flower, yellow, opening by pores or longitudinal slits; ovary inferior, of 2 or 3 (to 6) fused carpels, often angled to winged, styles separate or united basally, yellow, stigmas resembling stamens

FRUIT A dry winged capsule (rarely fleshy), with many tiny seeds

Although several genera have been described in Begoniaceae, just two are now recognized, of which only *Begonia* (for example, *B. undulata*, cane-stemmed begonia) is important in horticulture; the other genus is *Hillebrandia*. Mostly tropical, some *Begonia* species are hardy, dying back in winter and resprouting from an underground tuber. Many other species can of course be maintained indoors or in a cool greenhouse until warm weather returns or can simply be treated as annuals and replaced each year. *Begonia grandis* is hardy across most of the United States and parts of Canada, but in California, Florida, and warmer part of the South, several more species survive outdoors through most winters. Bedding begonias are hybrids developed for their compact habit and long flowering season. Likewise, tuberous begonias, those with large doubled flowers, are the product of breeding and selection by the nursery industry. Among the more attractive wild species is *B. boliviensis*, with large, fire-engine-red flowers borne on trailing stems; plants are best overwintered indoors. The orange-flowered

southern African *B. sutherlandii* is a low-growing species with orange blooms, hardy in areas with mild winters, and it too can be replaced each year from seedlings or will self-sow.

Many *Begonia* cultivars are available for indoor culture, most grown for the interestingly colored foliage. Among these are the rex begonias with large, somewhat hairy leaves in contrasting purple, red, green, and silver. There are many more available from nurseries, with new cultivars being produced annually by plant breeders.

Genera of Begoniaceae
Begonia • Hillebrandia

CUCURBITACEAE A largely tropical family, the cucumber family includes some 940 species and is closely related to Begoniaceae, also having unisexual flowers. Both male and female flowers are often borne on the same plant, the females larger and fewer than the males. Most Cucurbitaceae are vines and have prominent female flowers with a large inferior ovary and sterile stamens, and small male flowers with large anthers and lacking an ovary. All Cucurbitaceae are frost tender, and the family offers no truly ornamental plants but provides us with vegetables and fruit, including squash, marrow, and summer squash (*Cucurbita pepo*), pumpkin (*Cucurbita moschata*), cucumbers (*Cucumis sativus*), melons (*Cucumis melo*), and watermelon (*Citrullus lanatus*).

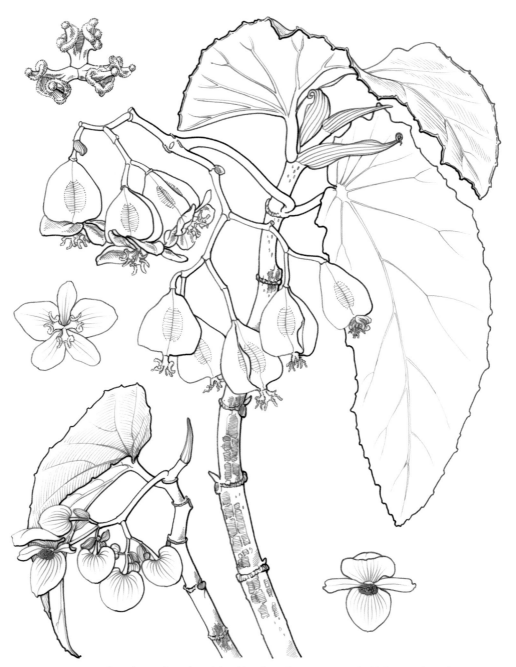

Begoniaceae, *Begonia undulata*. Many Begoniaceae are recognizable by conspicuously stipulate, asymmetric leaves, sometimes beautifully colored and marked, and by unisexual flowers borne in separate clusters on the same plant. Male flowers have two or four petal-like sepals and numerous stamens; female flowers usually have five or six petaloid sepals and a three-winged, inferior ovary with three intricately folded or twisted styles.

Begoniaceae, *Begonia sutherlandii*

BERBERIDACEAE

Barberry Family

About 600 species in 14 genera

RANGE mostly northern temperate, also tropical mountains, *Berberis* extending to southern South America

PLANT FORM Trees, shrubs, and perennial herbs, usually with yellow pigment (berberine) in tissues, especially roots; leaves simple or compound, then pinnate, bipinnate, ternate (in a cluster of three), or solitary with articulation, in spirals or basal (for example, *Podophyllum*), with minute stipules or without

FLOWERS Usually bisexual, radially symmetric, 2-, 3-, or 4-merous; in racemes, cymes, or solitary; calyx usually with 4 (but 3–8 or more) free sepals, green or petaloid and often soon falling; corolla with petals free, usually 8 in 2 whorls (sometimes many or in *Achlys* both sepals and petals lacking); stamens 4, 6, or many, anthers opening by valves (but not in *Nandina*, heavenly bamboo, nor *Podophyllum*); ovary superior, consisting of 1 carpel, usually with many ovules (2 in *Achlys*)

FRUIT Usually a berry or dry, then seeds often arillate

Including a diverse range of plants, from trees, woody shrubs, and perennials both soft and stiffly leathery, Berberidaceae include many attractive or unusual plants for temperate gardens. Spiny *Berberis* species, loosely called barberries, are often used as hedge plants, some cultivars with pale chartreuse or deep red foliage. *Berberis darwinii* has particularly striking orange flowers produced in abundance in spring, not the typical *Berberis* of hedges. It was discovered by Charles Darwin in southern South America on his epic voyage around the world. Unfortunately, *B. darwinii* is a noxious weed in parts of Australia and New Zealand. The North American and eastern Asian genus *Mahonia*, now included in *Berberis*, has several eastern Asian species with glossy and sometimes prickly foliage, many blooming in the winter months, a boon for temperate gardens. The native western North American mahonia (*B. aquifolium*) is invasive in western Europe. Berries of this and other *Berberis* species of the mahonia group are used for jams and jellies and were important in the diet of Native Americans.

The genus *Epimedium* includes some 55 species of temperate Eurasia and northern Africa, many of them worthy of horticultural attention. Flowers range from those with eight small petals in two whorls, not much different from flowers of *Berberis*, to others with quite large petals, those of the inner whorl bearing curious, small to quite prominent spur-like pouches. Numerous species and hybrids of *Epimedium* are available in the nursery trade today. The western North American *Vancouveria*, closely related to *Epimedium*, has quite small, perhaps less interesting, white or yellow flowers but is nevertheless a hardy and attractive ground cover. Seeds of

Berberidaceae, *Epimedium acuminatum*

Berberidaceae, *Epimedium* cultivar

both *Epimedium* and *Vancouveria* have soft white appendages (arils) attractive to ants, the agents of seed dispersal.

The strange growth form of *Podophyllum*, with large, umbrella-like leaves produced from an underground rootstock or short stocky stem, is an interesting accent plant for woodland gardens. Some species or cultivars of *Podophyllum* have quite extraordinary mottled foliage, including *P. delavayi*, *P. difforme*, and *P. hexandrum*. The flowers produced by these other-worldly plants are rather disappointing, borne below the few huge leaves. Flowers of some species, notably *P. pleianthum*, are unpleasantly to downright foul scented. The white-flowered, eastern North American and Midwestern *P. peltatum*, May apple, is so named for its edible fruit ripening at that time. The fruit is evidently not particularly tasty.

No discussion of Berberidaceae would be complete without mention of *Nandina*, the single species of which, *N. domestica* (heavenly bamboo), may have originated in central China but is now widely natural-ized in India and Japan. A handsome, evergreen shrub, it has clusters of bright red or sometimes yellow fruits, making it a useful ornamental in winter. The multiply pinnate leaves are unusual for the family.

Many Berberidaceae are used in traditional and modern herbal medicine, most notably *Epimedium*, which is important in Chinese medicine for vari-ous ailments, including hypertension. Extracts of *Podophyllum* species are used for treatment of vari-ous cancers. Some *Berberis* species have the dubious distinction of being the alternative hosts of rust fungi that cause disease in cereal crops.

Because the family includes such diverse types of plants, it is difficult to provide a simple rule for identifying Berberidaceae. Leaves are simple or compound, often of an unusually firm texture, and the flowers are often four-merous. The sepals are frequently inconspicuous and soon fall so are best examined in the bud state, and there are usually two whorls of petals, the inner of the two forming a cup and the outer outspread. In *Epimedium* the inner

Berberidaceae, *Podophyllum* hybrid

petals have spur-like extensions, these sometimes very prominent. Most genera have unusual valvate anthers, thus opening by terminal pores covered by flaps (valves). The valves close in inclement weather or even when flowers are shaken, a feature of just a few other plant families, notably the laurel family (Lauraceae). The ovary is apparently composed of a single carpel. Presence of the yellow pigment berberine in tissues of Berberidaceae is another distinctive feature of the family, also characteris-tic of some Papaveraceae and Ranunculaceae, both closely related families. Berberine has antibiotic and antifungal properties and has been found useful in treating some types of diabetes.

Selected Genera of Berberidaceae

Achlys • Berberis, including *Mahonia • Caulophyllum • Diphyllaea • Epimedium • Jeffersonia • Nandina • Podophyllum • Ranzania • Vancouveria*

BETULACEAE

Birch Family

About 140 species in 6 genera

RANGE Mostly northern hemisphere, also tropical mountains

PLANT FORM Deciduous trees and shrubs; leaves simple, in spirals, usually toothed, with stipules soon falling

FLOWERS Unisexual; males in elongate pendulous catkins; females in pendulous or erect catkins, or just a few flowers in a tight cluster; calyx of scale-like sepals or absent; corolla lacking; stamens as many as sepals or more, usually free; ovary inferior, of 2 or 3 united carpels with separate styles, 2- or 3-celled with 1 or 2 ovules per locule

FRUIT A nut subtended by prominent leafy bracts (as in filberts, *Corylus*) or dry with two wings (a samara), sometimes enclosed in a woody, cone-like structure

Betulaceae, *Carpinus caroliniana*, female inflorescences

Betulaceae, *Carpinus caroliniana*, male inflorescences

An important component of northern-hemisphere forests, Betulaceae are well known as ornamentals in gardens and street plantings. Like the closely related Fagaceae (oak family), Betulaceae have male flowers borne in drooping inflorescences called catkins, with reduced calyx and corolla. The stamens, exposed to the elements, produce light, aerodynamic pollen carried by the wind. Some birches (*Betula*) are grown in gardens for their papery white bark that stands out in both summer and winter; some Arctic species are no more than creeping shrubs. The genus has fruits carried in drooping cone-like structures that shatter when mature to yield samaras dispersed by wind. Wood of several species was used in past times for construction and furniture.

Several species of *Alnus* (alders) are valued for their timber. Distinctive in having their nuts enclosed in leafy bracts, species of *Carpinus* (hornbeam) are important in urban forestry, and the wood is still valued today for multiple specialized uses. *Corylus*, the filbert or hazel, is grown in orchards for the nuts used as food and in confectionary. The cultivar 'Contorta'

or corkscrew hazel with twisted stems is grown in gardens as a curiosity. *Corylus* catkins, produced in late winter or early spring, are often striking in leafless deciduous woodland. Pollen thought to represent Betulaceae is known from Cretaceous deposits more than 70 million years old. Among species of the family, Fang's or monkeytail hornbeam, *Carpinus fangiana*, native to China, has remarkable catkins, more than 16 inches (over 40 cm) long.

Betulaceae are recognizable by their small, simple leaves with serrated margins, male and female flowers on separate inflorescences, inconspicuous scale-like sepals, and the male flowers and sometimes the females borne in drooping catkins. The fruits are either small two-winged seeds or nuts enclosed in leafy bracts.

Genera of Betulaceae

Alnus • Betula • Carpinus • Corylus • Ostrya • Ostryopsis

Bignoniaceae, *Tecomaria capensis*. The opposite leaves of Bignoniaceae are frequently compound, typically with toothed leaflets. The elongate capsules (and winged seeds) are characteristic of the family, as are the tubular calyx and the tubular, two-lipped corolla. Most members have four stamens in two pairs, with strongly divergent anther lobes.

BIGNONIACEAE

Bignonia Family

About 860 species in 82 genera

RANGE Mainly tropics, especially Central and South America, also in temperate Eurasia, Africa, and North America

PLANT FORM Trees, shrubs, woody vines, and a few perennial herbs; leaves compound or simple, usually opposite or in whorls (rarely in spirals), without stipules

FLOWERS In terminal or axillary racemes, or panicles, perfect, large, and showy; calyx of 5 sepals united for most of their length; corolla with petals united in a tube, lobes often 2-lipped, with upper lip 2-lobed; stamens 4 in 2 pairs (or only 2), inserted on corolla tube, anthers with cells widely diverging, opening by longitudinal slits; ovary superior, of 2 united carpels, 1- or 2-celled, with many ovules, style 1, long, slender, 2-lobed at apex

FRUIT A pod-like capsule of distinctive appearance, usually like a large vanilla pod but typically containing numerous flat, winged seeds (rarely in the tropics, fruit fleshy)

With relatively few non-tropical species, Bignoniaceae are best known to North American gardeners through the large ornamental trees of southern and central North America, *Catalpa bignonioides* and *C. speciosa*, southern and northern catalpa (or catawba). Both are widely grown as street trees, sometimes in large gardens. The large white flowers speckled in the throat are borne in large panicles. Another member of the family used in temperate horticulture is the trumpet vine (*Campsis radicans*), native to southwestern North America, a popular garden subject, perhaps best regarded as half-hardy. The southwestern North American and Mexican genus *Chilopsis* (desert willow) has been crossed with *Catalpa;* the resulting hybrid, ×*Chitalpa*, is an attractive small tree. Several other vines of the family are grown in areas of Mediterranean climate and in the tropics. The southern African *Tecomaria capensis* (Cape honeysuckle) needs protection from frost but is hardy across southern North America, southern Europe, Australia, and New Zealand as well as southern Africa. Few other Bignoniaceae thrive in such climates. The Brazilian tree *Jacaranda mimosifolia* bears panicles of striking, large, blue flowers. In areas of light frost the flowers are produced before new leaves emerge, offering a spectacular sight. The only herbaceous member of the family is *Incarvillea*, a hardy perennial, most species of which are native to the mountains of China and Tibet. The large trumpet-like flowers are typical of the family. The name hardy gloxinia is loosely used for the genus or applied specifically to *I. sinensis* (gloxinias are otherwise members of the mostly tropical family Gesneriaceae, described below).

Among families with opposite leaves and a five-lobed corolla with petals united below into a cup or prominent tube, Bignoniaceae are recognizable by the woody

Bignoniaceae, *Podranea ricasoliana*

Bignoniaceae, *Markhamia obtusifolia*

habit (not *Incarvillea*), sometimes compound leaves, anthers with diverging lobes, ovary containing numerous ovules, and the characteristic fruits, which are dry capsules, usually elongate in shape with leathery to woody walls, the two halves separating from one another and releasing flattened, winged seeds.

The genus *Paulownia*, sometimes included in Bignoniaceae, Scrophulariaceae, or recognized as the separate family Paulowniaceae, is discussed under Scrophulariaceae.

Selected Genera of Bignoniaceae

Bignonia • Campsis • Catalpa • Chilopsis • Crescentia • Incarvillea • Jacaranda • Pandorea • Rhigozum • Spathodea • Tecoma • Tecomaria

GESNERIACEAE Allied to both Scrophulariaceae and Bignoniaceae, the Gesneriaceae (gloxinia family) are almost exclusively tropical and will not survive winters in temperate gardens. Many are grown indoors as pot plants, in heated greenhouses, or on the kitchen windowsill. These include *Saintpaulia* (African violet), native to Kenya and Tanzania, and the African and Madagascan *Streptocarpus*, sometimes called Cape primrose. Indeed, DNA data suggest that *Saintpaulia* is merely a specialized group of species within *Streptocarpus*. Species of the tropical American genera *Gloxinia* and *Sinningia* are often seen in greenhouses. Gesneriaceae include trees, shrubs, vines, and many perennial herbs, most with attractive flowers with tubular corollas. Leaves are simple, entire, and mostly opposite or whorled. Like Scrophulariaceae, the ovary is superior, consists of two united carpels, and the fruit is a capsule. The family is most readily distinguishable by the anthers usually joined in pairs (occasionally separate) and the one-celled ovary.

CALCEOLARIACEAE The genus *Calceolaria* (slipper flowers), striking in their flowers with the petals forming a conspicuous, inflated pouch, has historically been included in Scrophulariaceae or even Gesneriaceae, but current authorities consider it best treated as its own family. Several species are grown as ornamentals.

BORAGINACEAE

Borage or Forget-Me-Not Family

About 2600 species in 120 genera

RANGE Tropical and warm temperate, especially Eurasia

PLANT FORM Trees, shrubs, and herbs, including many perennials, some annuals; leaves simple, in spirals or rarely some opposite, usually entire, without stipules; stems and leaves sometimes completely smooth but usually hairy, often the hairs bristly and rough to the touch

FLOWERS Usually bisexual and radially symmetric, in distinctive coiled, terminal cymes (scorpioid cymes) coiled in early flowering, rarely flower solitary; calyx mostly 5-lobed (or 4- or 8-lobed), usually green; corolla with petals united basally or forming a cup or tube, mostly 5-lobed (or 4- or 6-lobed); stamens as many as petals, thus mainly 5, inserted on corolla tube; ovary superior, composed of 2 united carpels but often deeply lobed (not in *Heliotropium*), 2-celled but 4-lobed, with 1 terminal style, stigma 2-lobed or single, each carpel lobe usually with 1 ovule

FRUIT Usually comprising four single-seeded nutlets shed individually, sometimes barbed or bristly

Studies of the relationships of genera of flowering plants using DNA sequences show that the waterleaf family, Hydrophyllaceae (*Hydrophyllum*, *Phacelia*, and their close relatives), are nested in Boraginaceae, in which it is now included as the subfamily Hydrophylloideae. Most genera of horticultural interest are placed in subfamily Boraginoideae. Among these, *Cerinthe*, *Echium*, *Onosma*, and *Pulmonaria* (lungwort) have tubular, often nodding flowers, all of some horticultural interest, considerable in the case of *Pulmonaria*, for which there are many attractive selections, most with variously variegated leaves. Species of *Mertensia*, which also have tubular flowers, are useful hardy woodland perennials; Virginia bluebell, *M. virginica*, is the most common of the species in cultivation. The related genus *Symphytum* (comfrey) is an attractive, hardy ground cover. *Lithodora* is also occasionally grown in gardens for its intensely blue flowers but has uncomfortably bristly leaves and stems.

With typical forget-me-not flowers, *Anchusa*, *Cynoglossum*, *Myosotis*, *Omphalodes*, and several more genera have star-like, usually blue flowers with the petals united only near the base and white or yellow appendages at the base of the petals. Several genera in the tropics are trees, notably *Cordia* and *Ehretia*. The shrubby genus *Lobostemon* (for example, *L. fruticosus*, pajama bush) is an important element of the flora of western South Africa, and several species of the genus are useful ornamentals. Pride of Madeira, *Echium candicans*, with dense spires of blue flowers, is a good plant for mild climates and seaside gardens. The large genus *Heliotropium* includes the heliotrope of horticulture, *H. peruvianum*, grown as an annual for its deep blue, scented blooms.

Boraginaceae, *Lobostemon fruticosus.* Typical of many Boraginaceae
are the harsh bristles on the foliage and the five-lobed, bell-shaped
flowers in characteristically coiled cymes. The five stamens are inserted
on the corolla tube, which in *Lobostemon* have stiffly bearded protuber-
ances at the base of the filaments, and the ovary is deeply four-lobed.

Boraginaceae, *Onosma fruticosa*

Boraginaceae, *Lithospermum purpurocaeruleum*

Boraginaceae, *Omphalodes cappadocica*

Boraginaceae, *Cerinthe major*

Boraginaceae, *Borago officinalis*

Boraginaceae subfamily Hydrophylloideae also include several annuals and perennials worthy of horticultural attention. Several species of *Phacelia* provide striking spring displays across North America, from the woodland of the East Coast to the deserts of California and Arizona. The East Coast native *P. fimbriata*, with its fringed petals, is one of the most attractive species of the genus. Species of *Nemophila* (baby blue-eyes) are annuals of woodland and grassland and among the loveliest members of the California flora. Several species are grown in gardens devoted to native flora, and some more widely. *Nemophila menziesii* has lovely sky blue flowers with a white center.

While members of Boraginaceae are often distinctively hairy, even bristly and rough to the touch, a significant number of species have quite smooth leaves and stems, notably the genus *Cerinthe*. The flowers of borage (*Borago officinalis*) are used in salads. Lungwort (*Pulmonaria officinalis*) was so named for the speckled leaves said to resemble

diseased lungs, not for curing pneumonia. Species of comfrey (*Symphytum*) are rich in allantoin, which promotes healing of connective tissue and bone, used externally. Comfrey tablets and tea are said to cause liver failure when taken liberally. Locally, species of *Heliotropium* are used medicinally as purgatives and to control fertility.

Selected Genera of Boraginaceae

Amsinckia • Anchusa • Borago • Cerinthe • Cynoglossum • Echium • Eritrichum • Heliotropium • Hydrophyllum • Lithodora • Lobostemon • Mertensia • Myosotis • Nama • Nemophila • Omphalodes • Onosma • Phacelia • Pulmonaria • Symphytum • Trachystemon

BRASSICACEAE

Mustard Family

About 3500 species in 340 genera

RANGE Cosmopolitan, very diverse in the Mediterranean, western Asia, western North America, and the Andes of South America, few in tropics

PLANT FORM Mostly herbs, either perennial and annual, a few shrubs, rarely with woody stems; leaves simple, sometimes lobed to pinnately dissected, rarely pinnately or palmately compound, in spirals (rarely opposite), sometimes in basal rosettes, without stipules; always with mustard oils, hence with sharp peppery to mustard taste

FLOWERS Usually bisexual, radially symmetric, usually in bractless spikes or racemes, rarely solitary; calyx of 4 sepals, usually free; corolla of 4 free petals, usually with narrow, elongate base (claw), usually yellow to white, sometimes blue, pink, or mauve, rarely petals lacking; stamens mostly 6 with 2 in outer whorl shorter, sometimes only 4 or 2; ovary superior, of 2 united carpels with 2 styles, 2-celled, mostly with many ovules in 2 rows separated by a septum

FRUIT Dry capsules of characteristic appearance with walls falling to reveal interior septum to which seeds are attached (a silique or, if short, a silicule), smooth or constricted or "beaded" between seeds, sometimes fruit one-seeded and not splitting open

The traditional family name, Cruciferae, alludes to the cross-shaped flowers with four petals in opposed pairs. Brassicaceae include a fair number of garden plants, mostly low-growing perennials or subshrubs, including basket of gold (*Aurinia saxatilis*), candytuft (*Iberis*), wallflowers (*Erysimum*, formerly *Cheiranthus*), sweet alyssum (*Lobularia*), and honesty (*Lunaria*). Several species of *Cardamine* are grown as ground covers, among them white-flowered *C. trifolia*. Many more genera are grown as annuals, of which sweetly scented stocks (*Matthiola*) are actually short-lived perennials. Several annual species of the southern African genus *Heliophila* have unusual (for the family) deep blue flowers that open only in sunlight.

In addition to garden plants, several members of the genus *Brassica* are agricultural crops, including brussels sprouts, cabbage, cauliflower, broccoli, and kale, all selected cultivars of *B. oleracea* or hybrids with that species and mustard, *B. nigra*. Canola oil, also known as rapeseed oil, is derived from seeds a cultivar of *B. napus*. Radishes are domesticates of the genus *Raphanus*. Rocket (arugula or roquette) is a domesticate of *Eruca vesicaria*. Watercress is a species of the genus *Nasturtium* (not to be confused with the garden nasturtium, the genus *Tropaeolum* of the family Tropaeolaceae; see the appendix, Genera of Small Families Otherwise Not in General Cultivation). Many more species of Brassicaceae are grown locally in other parts of the world for greens. Horseradish (*Armoracia*) and wasabi (*Eutrema wasabi*) add to the list of plants cultivated for human consumption; their sharp to pungent taste is among the strongest in the family. Unfortunately,

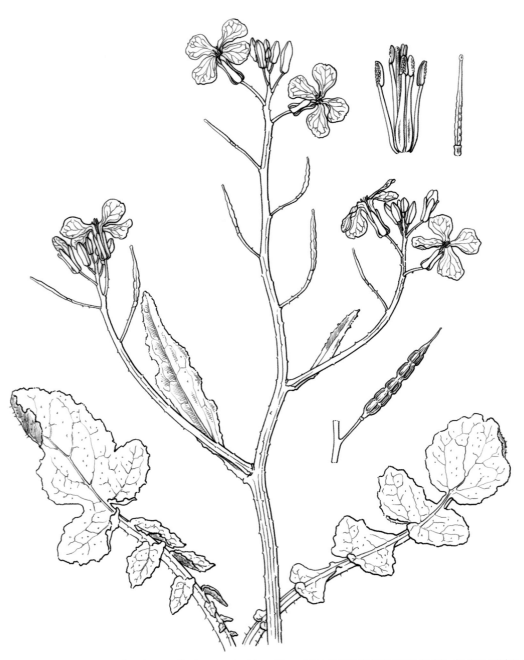

Brassicaceae, *Raphanus raphanistrum*. The rosette of raggedly pinnatisect (margin cut down to the midrib) leaves in wild radish is typical of many Brassicaceae, but the diagnostic characteristics for the family are the four free sepals in two pairs, these rapidly deciduous, the four free petals with long, narrow claws, and the six free stamens in pairs with one pair shorter than the others. The fruit in the family is usually dehiscent, but in *Raphanus* the seeds remain enclosed in the ribbed, bead-like segments. All Brassicaceae contain mustard oils, imparting to them a characteristic odor when bruised.

Brassicaceae, *Heliophila africana*

Brassicaceae, *Ricotia lunaria*

Brassicaceae, *Aubrieta deltoidea*

some Brassicaceae are significant weeds of agricultural lands, some invasive. The insignificant annual *Arabidopsis thaliana* is an important laboratory subject, used to study genetics and development in plants. Its short life cycle, just a few weeks, and small genome render it invaluable for such studies.

The mustard oils of Brassicaceae are glycosides and in concentrated form are toxic to humans, causing goiter. Glycosides have important antibacterial and medicinal properties and defend the plants against predation, especially by insects and, often, grazing mammals. The cabbage white butterfly caterpillar, a significant insect pest to crops of the family, tolerates mustard oils, metabolizing them into nontoxic substances. *Isatis tinctoria* was cultivated in antiquity for the blue dye, woad, obtained from the dried, fermented leaves. Woad was used by Celtic warriors to paint their bodies in preparation for battle.

Brassicaceae are readily identified by the cross-shaped flowers with four petals, usually six stamens (or four or two), and ovary of two united carpels with two styles. A taste test often confirms identification; the leaves and stems are peppery or have a mustard-like bite, sometimes very bitter. Despite the sharp to nasty taste, species of Brassicaceae are not poisonous to humans.

Selected Genera of Brassicaceae

Aethionema • Alyssum • Arabidopsis • Arabis • Armoracia • Aubrieta • Aurinia • Brassica • Cakile • Cardamine • Crambe • Draba • Eruca • Erucastrum • Erysimum • Eutrema, including *Wasabia • Heliophila • Iberis • Isatis • Lepidium • Lunaria • Malcolmia • Matthiola • Nasturtium • Raphanus • Sinapis • Thlaspi*

Capparaceae, *Capparis spinosa*

CAPPARACEAE The plants in this family are sometimes included in Brassicaceae, an action that radically changes the circumscription of the mustard family. Few Capparaceae are in cultivation except caper (*Capparis spinosa*) and *Cleome* (though the most current taxonomy places the latter in a separate family, Cleomaceae. Like Brassicaceae, Capparaceae have a calyx and corolla of four members, but six or many more stamens (six or fewer in Brassicaceae). Capparaceae typically have the ovary borne on a short to long stalk, a gynophore. The fruits, either capsules, berries, or drupes, also differ from the characteristic siliques or silicules of Brassicaceae. *Cleome* species are attractive annuals with compound leaves, and some are grown in gardens and parks. The flower buds of *Capparis spinosa* and some other species are pickled as capers, and the green fruits as caper berries, used as a flavoring in cooking.

BROMELIACEAE

Pineapple Family

About 2975 species in 60 genera

RANGE Mostly Neotropical region, a few subtropical, one genus in western Africa

PLANT FORM Terrestrial xerophytic (adapted to dry habitats) perennials or shrubs, some with thick stems and somewhat tree-like, and many epiphytes; leaves simple, often tough in texture, in spirals or clustered in rosettes, margins often serrate to spiny or reduced to scales, without stipules

FLOWERS Usually bisexual, radially symmetric or slightly asymmetric, usually in racemes, spikes, or heads, often with prominent, colored bracts more conspicuous than flowers; calyx of 3 free or basally united sepals, either green or colored and petaloid; corolla with 3 free or basally united petals; stamens 6 in 2 whorls, sometimes partly united, anthers opening by longitudinal slits; ovary superior or inferior, compound, of 3 united carpels, 3-celled with few to many ovules, with 1 terminal style usually 3-forked at apex

FRUIT Usually dry capsules, splitting completely, or a berry, rarely a compound fruit (for example, pineapple, *Ananas comosus*); seeds in capsules are often winged or plumose

An important family in the Neotropical region, Bromeliaceae include many species grown as houseplants and for display in greenhouses. Genera in cultivation include many *Bromelia* species, *Neoregelia*, and *Nidularium*. The genus *Tillandsia* (more than 450 species) includes epiphytes, the best known of which is *T. usneoides* (Spanish moss); plants resemble a lichen and drape on branches of trees and shrubs in humid southeastern North America. It is grown widely where it will survive winter conditions as a curiosity; the attractive deep blue flowers are seldom seen. The plants were formerly dried and used a packing material or stuffing and even fiber. Several other Bromeliaceae are used for fiber. *Ananas comosus*, the pineapple, is widely grown for its fruit, a compound structure developing from the entire inflorescence, the individual flower parts remaining as dry scales on the surface.

Selected Genera of Bromeliaceae
Ananas • Bromelia • Neoregelia • Nidularium • Puya • Vriesia • Tillandsia

BUXACEAE

Boxwood Family

About 115 species in 7 genera

RANGE Almost cosmopolitan but mainly Old World, few in North or South America

PLANT FORM Evergreen trees and shrubs or perennial herbs; leaves simple, in spirals or opposite (*Buxus*), without stipules

FLOWERS Bisexual or unisexual, radially symmetric, in small heads or spikes, often sweetly scented; perianth of 2 whorls of 2 or 3 units, often small, calyx and corolla usually not differentiated, tepals green, white, or pink; when unisexual, male flowers larger and more numerous; stamens usually 4 or many, anthers with longitudinal slits; ovary superior, mostly of 3 (or 4) united carpels with as many cells, each with (1 or) 2 ovules, styles as many as carpels, thus 3 or 4

FRUIT A drupe, or capsule, sometimes splitting explosively

Buxaceae, *Sarcococca hookeri*

Buxaceae are known to gardeners in the temperate zone mainly as the hardy boxwoods (*Buxus*) and winter-flowering sweet box (*Sarcococca*). Most Buxaceae are evergreen shrubs or trees, mostly with bisexual flowers. Flowers of *Buxus* and *Sarcococca* are, however, unisexual with the males larger; female flowers are usually borne in the same inflorescence or nearby; in *Buxus* several male flowers surround one or a few female flowers. True boxwoods have opposite leaves, distinctive in their small size, shape, and venation and are often used as hedges or topiary, especially the numerous cultivars of *B. sempervirens* (European or common box) and *B. microphylla* (Japanese box), the latter used an ornamental in Japan since at least 1450. *Buxus sempervirens* has been grown as an ornamental at least since Roman times in Europe, perhaps even longer than that.

The hardy, winter-flowering genus *Sarcococca* is valuable in landscaping in cold climates. The white or rarely pink-flushed flowers are strongly scented and, although small, make a modest display in months when little else is in bloom. With fleshy fruit, *Sarcococca* is spread by birds, and new plants spring up in surprising places. The genus *Pachysandra* has one species in eastern North America, *P. procumbens* (Allegheny spurge), and two or three in eastern Asia. *Pachysandra terminalis* (Japanese spurge) is the only Asian species widely cultivated and is used as a hardy ground cover, especially for shaded areas. The pachysandras were thought to be related to *Euphorbia*, hence the use of "spurge" in the common names. The relationships of Buxaceae are uncertain despite extensive anatomical, chemical, and molecular study. One suggestion is that they are distantly related to Pittosporaceae. *Buxus* provides a fine-grained hardwood used in inlays and musical instruments; in antiquity it was used for boxes, combs, furniture, flutes, etc.

Genera of Buxaceae

Buxus • Didymeles • Haptanthus • Notobuxus • Pachysandra • Sarcococca • Styloceras

SIMMONDSIACEAE *Simmondsia chinensis*, the only species in the genus and a misnomer as it is native to southwestern North America, was historically included in Buxaceae but is now known to be unrelated to the boxwoods and has been assigned to its own family. Jojoba is a drought-tolerant shrub or small tree. The nut-like, hard seeds are the source of jojoba oil, used in cosmetics and as a lubricant.

CACTACEAE

Cactus Family

About 1210 species in 115 genera

RANGE Almost exclusively in North and South America, mostly in arid habitats, one species in western Africa

PLANT FORM Mostly stem succulents, including thick-stemmed tree-like forms with few branches, or forming almost stemless rosettes, also fleshy shrubs, sometimes epiphytes; stems sometimes constricted or jointed into sections, often with spines, bristles, or hairs in tufts at nodes (called areoles), the spines, bristles, etc. derived from axillary buds, short shoots, or bud scales; leaves simple, in spirals, usually small and ephemeral or absent but present and succulent in a few genera

FLOWERS Usually bisexual, radially symmetric, large, usually solitary, borne at spine clusters or branch tips; perianth of many tepals in spirals, not differentiated into calyx and corolla, united basally and often fleshy; stamens many in spirals or clusters, anthers with longitudinal slits; ovary inferior, of 3 to many united carpels, with 1 style with as many arms as carpels, ovules usually many in 1 locule

FRUIT Usually a many-seeded berry (for example, prickly pear, *Opuntia*), often leathery and with spines or bristles, rarely dry and capsule-like; seeds often with fleshy appendages

Although well known to gardeners and the public at large, few Cactaceae are hardy in temperate gardens. Members of the family thrive, however, in greenhouses and indoors as pot plants. In southwestern North America several Cactaceae are grown in gardens and many more by cactus enthusiasts. The Christmas cactus (*Schlumbergera*) belongs to a small genus of eastern Brazil and has stems consisting of several flattened, leaf-like segments. Easily grown indoors, plants bloom in the winter months, providing a colorful and reliable display. The statuesque and emblematic saguaro (*Carnegia gigantea*) is a distinctive member of the flora of the Sonoran Desert of Mexico, Arizona, and New Mexico.

Cactaceae offer few examples of edible or useful plants, but the prickly pears (*Opuntia*) produce large edible fruits, sometimes seen in markets. When crushed, the fruit yields a sweet, refreshing juice. The red or pink dye, cochineal, is obtained from a scale insect that infests some *Opuntia* species. Several species of *Opuntia* have become aggressive invaders of the native flora in dry areas of the world, and the cochineal insect has been introduced in some places in the hope of controlling or even eliminating unwanted plants.

Cactaceae are usually recognizable among succulent plants by the absence of true leaves, at least at maturity (although the fleshy stem segments of some species may look leaf-like), usually with clusters of bristles, spikes, and hairs, sometimes even

Cactaceae, *Opuntia polyacantha*

Cactaceae, *Echinocereus triglochidiatus*

woolly, at the nodes. A few genera have large succulent leaves on mature plants, notably *Pereskiopsis*. Flowers have many petal-like segments united basally but not differentiated into calyx and corolla, many stamens, an inferior ovary consisting of a single locule with numerous ovules, and a style divided into several branches at the tip.

Selected Genera of Cactaceae

Carnegia • Cereus • Echinocereus • Epiphyllum • Hylocereus • Mammillaria • Melocactus • Opuntia • Pereskia • Pereskiopsis • Rhipsalis • Schlumbergera • Selenocereus

CAMPANULACEAE

Bellflower or Lobelia Family

About 2250 species in 55 genera

RANGE Tropical and warm temperate, especially northern hemisphere and southern Africa, also tropical mountains; especially diverse in the Mediterranean and western Asia and in southern Africa

PLANT FORM Mostly herbs, many perennials and some annuals, a few shrubs, pachycaul trees, sometimes twining (many *Cyphia* species), with tubers in *Cyphia;* leaves simple, in spirals, rarely opposite or in whorls, sometimes in a basal rosette, without stipules; with milky latex

FLOWERS Usually bisexual, radially symmetric or zygomorphic, usually in racemes, sometimes in cymes; calyx mostly 5-lobed, usually green, persisting in fruit; corolla with petals partly united (rarely free), mostly 5-lobed; stamens usually 5 (or as many as 10), grouped together, separating after pollen release, or united, usually inserted at base of corolla; ovary inferior, compound, of 1–5 but usually 3 united carpels with as many cells as carpels, with 1 terminal style, stigma capitate and emerging through tube formed by contiguous anthers or with as many lobes as carpels and then stamens withering in bud after depositing pollen on the style, each carpel with many ovules

FRUIT Usually dry capsules, splitting completely or partially through short slits at the top (rarely a berry), usually with many tiny seeds

Campanulaceae include two large and horticulturally important subfamilies, Campanuloideae and Lobelioideae, the latter sometimes recognized as a separate family (Lobeliaceae). In flowers of *Campanula* and its allies, pollen from the loosely united anthers is shed in the bud onto the outside of the stigma itself; after the flowers open the stigma lobes then unfold, concealing the pollen beneath them. Many *Campanula* species are striking ornamentals, many fully hardy, especially *C. persicifolia*, a common garden plant, *C. rapunculus* (rampion), and *C. rapunculoides* (creeping rampion). The latter spreads by slender, creeping rhizomes that easily break and develop into new plants; as a result the species tends to be weedy and is difficult to eradicate. *Campanula rotundifolia*, harebell or bluebell, with smaller but nevertheless attractive flowers, has a pan-northern-temperate distribution. The old-fashioned garden plant *C. medium*, Canterbury bells, is a biennial with the largest flowers in the genus. *Platycodon grandiflora*, the balloon flower, is an eastern Asian native often grown in temperate gardens. Some *Phyteuma* species (horned rampion) are occasionally grown as ornamentals in alpine and rock gardens.

Flowers of subfamily Lobelioideae are actually resupinate (twisted though 180 degrees), and the anthers are joined to one another. The style emerges through

Campanulaceae, *Wahlenbergia melanops*. Flowers of many Campanulaceae are notable for the way in which pollination is effected. Anthers shed their pollen directly onto the underside of the stigma lobes while still in bud and are visible only as withered remains after the flower opens. The style acts as a pollen presenter until the stigma lobes spread open to expose their receptive surface. The five petals are united into a bell- or salver-shaped corolla in *Wahlenbergia*.

Campanulaceae, *Campanula rupestris*

Campanulaceae subfamily Lobelioideae, *Michauxia campanuloides*

Campanulaceae, *Campanula*

the anther tube, pushing pollen out of the flower as the stigma lobe unfolds. In *Lobelia* the anthers bear a tuft of terminal hairs, and the corolla tube is cleft down one side. Some *Lobelia* species are tree-like, usually singled stemmed, with leaves and flowers crowded at the crown as in the giant lobelias of the eastern African mountains. Although lobelias are best known in gardens as low, tufted, or trailing herbs, a range of conspicuous species with red, yellow, or bicolored flowers occurs in Central and South America, some half-hardy and grown in warm temperate gardens. Hummingbird-pollinated *Lobelia* species in North and tropical America have red or yellow flowers. Among these, red-flowered *L. tupa*, native to Chile and sometimes called devil's tobacco (extracts are a nicotine agonist), is a distinctive perennial species that is becoming popular as an ornamental. Native North American species include red-flowered *L. cardinalis* (cardinal flower) and blue-flowered *L. siphilitica* (great blue lobelia), both used medicinally by Native Americans, especially as an emetic. The use of extracts for treatment of venereal disease was largely ineffective but is the inspiration for the specific name *siphilitica*.

Campanulaceae can be recognized by the presence of milky latex when leaves or stems are broken, frequently blue flowers (not so in the Americas), stamens with anthers either connate or deciduous, an inferior ovary, and tubular (*Lobelia* and relatives) or bell-shaped corolla (*Campanula, Platycodon, Wahlenbergia*, including *W. melanops*, dark-eyed wahlenbergia).

Selected Genera of Campanulaceae

Burmeistera • Campanula • Centropogon • Cyanea • Cyphia • Downingia • Lobelia • Monopsis • Phyteuma • Platycodon • Prismatocarpus • Roella • Wahlenbergia

Cannabaceae, *Celtis sinensis*. The small flowers of *Celtis* are unisexual and radially symmetric, the males with five free stamens and the females with a superior ovary bearing a solitary, deeply two-lobed style. The asymmetric leaves with three main veins from the base resemble those of Ulmaceae (elms and relatives).

CANNABACEAE

Cannabis or Hemp Family

About 90 species in 9 genera

RANGE Mostly northern temperate and Mediterranean, also African highlands to South Africa

PLANT FORM Trees, shrubs, vines, perennial and some annual herbs; leaves opposite or in spirals, usually simple, often toothed, evidently palmately lobed or digitately compound (*Cannabis*), stipules usually present and persistent

FLOWERS Usually bisexual or unisexual, radially symmetric, in cymes, or female flowers crowded together; calyx of 5 sepals; corolla lacking; stamens usually 5; ovary superior, compound, of 2 united carpels with 1 cell and 1 ovule, style either 1 but divided to middle or 2 separate styles

FRUIT A drupe or achene covered by the calyx

The circumscription of Cannabaceae has historically been particularly unsettled. Molecular studies, however, show that the family not only includes *Cannabis* (hemp) and the vine *Humulus* (the genus of hops) but also trees, including *Celtis* (for example, *C. sinensis*, Chinese hackberry) and *Trema*, both historically usually included in Ulmaceae (described below). *Cannabis*, of course, is the source of marijuana (ganja, hash, hashish, kef), the dried leaves and flowers or gum of which can be smoked (or eaten) for their intoxicating, stimulating, and sometimes analgesic and medicinal effects. Laws in many countries have outlawed the cultivation of *Cannabis* because of its use as a drug, blithely ignoring its value as a fiber, thus replacing this use in the United States in the 19th century by the expanded cultivation of cotton, which is subsidized at great cost. Up to the late 19th century most paper in the world was made from cannabis hemp fiber, including such famous books as the Gutenberg and St. James Bibles and the first drafts of the Declaration of Independence. Today, *Cannabis* continues to provide a less expensive source of fiber (hence the common name hemp) for clothing, bags, canvas, paper, and much more. The cultivation of strains of *Cannabis* high in fiber and low in intoxicating compounds has increased. Ground seeds of fiber-producing strains, low in physiologically active cannabinoid compounds, provide a nutritious spread resembling peanut butter in parts of eastern Europe. *Cannabis*, which in most taxonomic accounts includes just one species, *C. sativa*, is readily recognizable in the family by the digitately compound leaves, the leaflets with serrated margins.

Rather different in habit and leaves, *Humulus* is a perennial vine with palmately lobed leaves, the fruits of which are called hops, used primarily to flavor and stabilize beer. Thousands of acres are devoted to hop production, and even more to that of *Cannabis*, the latter sometimes illegally. Other members of Cannabaceae are rarely grown in gardens, but *Celtis* species (hackberries or nettle trees) are

Cannabaceae, *Cannabis sativa*

grown in parks and as street plantings and sometimes in gardens, particularly in southern Africa, where *C. africana* and *C. sinensis* are valued for their attractive appearance, rapid growth, and tolerance for hot, dry conditions and low rainfall. Arborescent genera of Cannabaceae also provide useful timber, especially *Celtis*, *Chaetacme*, and *Trema*.

Basic floral structure in Cannabaceae is unspecialized; the small flowers are bisexual or unisexual and have a perianth of a single whorl, usually referred to as a calyx. Male flowers have five stamens and females a superior ovary with a single divided style or two styles.

Selected Genera of Cannabaceae
Cannabis • Celtis • Chaetacme • Humulus • Trema

ULMACEAE The Ulmaceae, in which several genera of Cannabaceae were included in the past, now include *Ulmus* (elms), *Zelkova* (Japanese elm or zelkova), and a handful of other genera not in general cultivation. Like Cannabaceae, these woody plants have similarly reduced flowers but dry, winged fruits. Both *Ulmus* and *Zelkova* are important in urban forestry and for timber production. Other genera of Ulmaceae are also used for timber. Most Ulmaceae have leaves asymmetric at the base, a useful aid to identification but to be used with caution as this characteristic is not unique to Ulmaceae.

CAPRIFOLIACEAE

Honeysuckle Family

About 900 species in 39 genera

RANGE Mostly northern temperate and Mediterranean, also African highlands to South Africa

PLANT FORM Small trees, shrubs, vines, perennials, and some annuals; leaves usually simple, often toothed, or somewhat dissected, opposite or whorled, stipules small or usually absent

FLOWERS Usually bisexual, radially symmetric or zygomorphic, in cymes, flat-topped to rounded clusters, or heads; calyx 4- or 5-lobed, sepals united below or reduced to bristles or lacking, usually green; corolla with petals partly united, mostly 5-lobed (or 3- or 4-lobed); stamens usually 5 (sometimes fewer, only 1 in *Centranthus ruber*), inserted on corolla tube and alternating with corolla lobes; ovary inferior, compound, of (2 to) 5 united carpels with as many cells as carpels, with 1 terminal style, each carpel with 1 to many ovules

FRUIT Usually drupes or berries, sometimes dry capsules

The family Caprifoliaceae have expanded as a result of molecular studies showing that the herbaceous genera usually regarded as the separate families Dipsacaceae (including *Cephalaria rigida*, for example) and Valerianaceae are nested with the larger woody genera with which they share similar to identical floral structure. These same studies show that *Sambucus* (elderberry) and *Viburnum* are allied to the small, perennial genus *Adoxa*, and these three genera are now included in the family Viburnaceae (described below).

Caprifoliaceae have contributed many species and cultivars to horticulture. Most common in gardens are *Abelia*, *Weigela*, and the honeysuckles (*Lonicera*). This last is a genus of about 180 species of evergreen and deciduous shrubs and vines, including the common honeysuckle, *L. periclymenum*, also called woodbine, with sweetly scented white flowers fading to creamy yellow. Less often grown in gardens are *Dipelta*, *Heptacodium*, *Kolkwitzia*, and *Leycesteria*. Among the perennials are species of *Dipsacus*, *Knautia*, *Scabiosa*, and *Valeriana*, in which the flowers are aggregated in rounded clusters to flat heads often surrounded by conspicuous and sometimes spiny bracts. Often grown as an annual, *Centranthus ruber* (red valerian or Jupiter's beard) has clusters of small red, pink, or white flowers with just a single stamen, unusual for Caprifoliaceae. The spiny, dried heads of the Eurasian *Dipsacus*, teasel, were used in past times to brush or tease cloth, and *D. sativus* has become a common weed across North America. *Valerianella* is used in salads (mâche or lamb's lettuce). Tincture of valerian, used to induce sleep, is derived from species of *Valeriana*.

Caprifoliaceae can be recognized by the simple (sometimes dissected), opposite leaves, inferior ovary, flowers with the corolla lobes united below, five or fewer stamens, and cymose or capitate inflorescence. Viburnaceae (see below) are

Caprifoliaceae, *Lonicera periclymenum*. A twining vine, this honeysuckle has zygomorphic flowers in umbel-like clusters, unusual for the family, but is otherwise typical in its opposite leaves lacking stipules, five-lobed calyx, tubular corolla, five stamens, and single style.

Caprifoliaceae, *Scabiosa incisa*

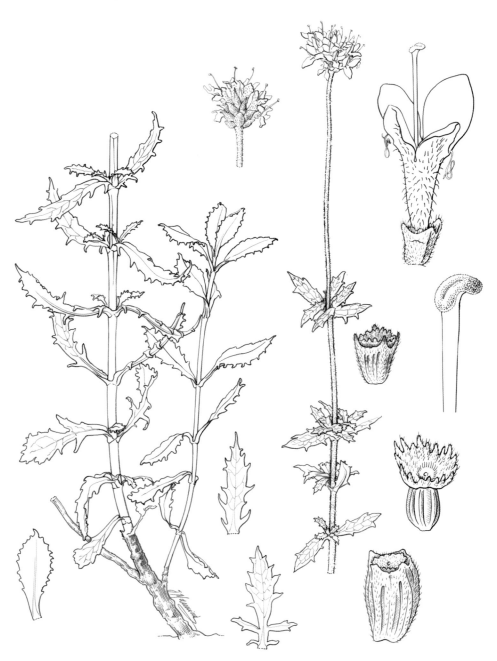

Caprifoliaceae, *Cephalaria rigida. Cephalaria* and related genera are typical of the herbaceous members of Caprifoliaceae in their opposite, toothed or lobed leaves lacking stipules, and heads of small flowers with the outermost more strongly two-lipped. Individual florets are funnel-shaped with four unequal lobes and a finely toothed, cup-shaped calyx. The inferior ovary is enclosed within a toothed involucre, and the stigma is unlobed; the involucre persists in fruit.

Caprifoliaceae, *Lonicera periclymenum*

Caprifoliaceae, *Kolkwitzia amabilis*

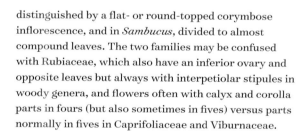

Viburnaceae, *Viburnum burkwoodii*

Viburnaceae, *Viburnum dentatum*

distinguished by a flat- or round-topped corymbose inflorescence, and in *Sambucus*, divided to almost compound leaves. The two families may be confused with Rubiaceae, which also have an inferior ovary and opposite leaves but always with interpetiolar stipules in woody genera, and flowers often with calyx and corolla parts in fours (but also sometimes in fives) versus parts normally in fives in Caprifoliaceae and Viburnaceae.

Selected Genera of Caprifoliaceae
Abelia • Centranthus • Dipelta • Dipsacus • Heptacodium • Knautia • Kolkwitzia • Leycesteria • Linnaea • Lonicera • Scabiosa • Symphoricarpos • Valeriana • Valerianella • Weigela

VIBURNACEAE A family of four genera and some 225 species immediately allied to Caprifoliaceae, the Viburnaceae include two fairly widely cultivated genera, *Sambucus* and *Viburnum*. The latter, a widespread genus of 210 species, has contributed several ornamental shrubs to gardens, notably the Mediterranean *V.*

tinus, *V. macrocephalum* (snowball tree), *V. carlesii*, and *V. tomentosum*. The winter-flowering and fragrant *V. ×bodnantense* 'Pink Dawn' is a hybrid between winter-flowering *V. farreri*, with white fragrant flowers, and *V. grandiflorum*. The other horticulturally important genus of Viburnaceae is the elderberry, *Sambucus*.

The berries of *Viburnum*, often blue to purple, are sometimes edible, but some are poisonous, notably those of *V. tinus*. Fruits of *Sambucus nigra* are used to make elderberry wine. The cultivar *S. nigra* 'Black Lace', with dark red, much-divided foliage, is a striking large shrub for gardens. The remaining two genera of Viburnaceae are perennial herbs of no horticultural value. For convenience, Viburnaceae may be included in Caprifoliaceae; the differences between them seem trivial.

Genera of Viburnaceae
Adoxa • Sambucus • Sinadoxa • Viburnum

Caryophyllaceae, *Silene vulgaris*. Like most Caryophyllaceae, species of *Silene* have opposite leaves and thickened nodes. The flowers of campions have petals with small lobes or scales at the base of the limb forming a corona, and a tubular calyx partly enclosing the claw-like bases of the five petals and ten stamens. The ovary has five separate, thread-like styles, and the kidney-shaped seeds are often beautifully ornamented.

CARYOPHYLLACEAE

Carnation Family

About 2500 species in 96 genera

RANGE Cosmopolitan but especially diverse in temperate and warm northern hemisphere

PLANT FORM Mostly shrubs, perennial herbs, and annuals, rarely small trees or twining plants, often with swollen nodes; leaves opposite (rarely in spirals), simple, entire, usually without stipules

FLOWERS Mostly bisexual, radially symmetric, in cymes or solitary; calyx of 4 or 5 united sepals, often persistent around fruit (or sepals free); corolla of 4 or 5 free petals, often clawed with large expanded limb and forked at tips (petals lacking in some genera not in cultivation); stamens usually 5 or 10, inserted at bases of tepals, anthers with longitudinal slits; ovary superior, of 2–5 united carpels, 1-celled (rarely partially segmented below middle) with several to many ovules, styles as many as carpels or united into 1, short or long

FRUIT A capsule, rarely a berry or nutlet

Caryophyllaceae include many attractive perennials and not a few annuals, several suitable for rock and alpine gardens. Perhaps most important is the genus *Dianthus*, which includes carnations and pinks, many prized for their scented flowers. Long grown in European gardens, *Dianthus* species and cultivars are now grown in cool and warm temperate gardens throughout the world. *Gypsophila* is another old garden favorite, perhaps more so in the past than today. Known as baby's breath, the small pink or white flowers produced in profusion make an attractive display and excellent cut flower, frequently used by florists. Species of *Silene* are also grown in temperate gardens. Species with glandular stems, leaves, and calyces are known as catchflies and do indeed trap small insects but merely incidentally, while species with smooth stems and leaves are commonly called campions. *Silene* (for example, *S. vulgaris*, bladder campion) is distinctive in having small appendages (a corona) at the base of the petal limbs.

Species of the genera *Lychnis* and *Melandrium* are now included in *Silene*, rendering it one of the largest genera of the family with some 700 species. *Silene* (*Lychnis*) *coronaria* (rose campion) is widely cultivated and naturalized in many places, and *S.* (*Lychnis*) *chalcedonica* is the Maltese cross. Species of *Saponaria* (soapwort) also provide ornamentals, notably the trailing *S. ocymoides* in rock gardens. The generic name (Latin *sapo*, soap) celebrates its use in the past as a gentle soap made by boiling up crushed leaves or roots.

Delightful as many Caryophyllaceae are to gardeners, others are vexing weeds. Both *Cerastium* and *Stellaria* are known as chickweed. *Stellaria media* is edible as a salad green and also has medicinal properties. The fine-leaved, procumbent plant

Caryophyllaceae, *Dianthus angulatus*

Caryophyllaceae, *Dianthus orientalis*

Caryophyllaceae, *Silene undulata*

Caryophyllaceae, *Gypsophila elegans*

Caryophyllaceae, *Silene*

Sagina procumbens (pearlwort) is also weedy and particularly difficult to eradicate in garden situations. Extremely hardy, it is sometimes grown in gardens in North America where it is sold as English moss!

Caryophyllaceae are usually easy to recognize by their jointed stems with swollen nodes, opposite, entire leaves, and four- or more often five-merous flowers, usually with a calyx of united sepals enclosing the corolla of free, often bilobed or fringed petals. The fruit is usually a capsule, with a single locule containing seeds attached to a central column.

Selected Genera of Caryophyllaceae
Arenaria • Cerastium • Dianthus • Drymaria • Gypsophila • Sagina • Saponaria • Silene, including *Lychnis, Melandrium,* and *Viscaria • Spergularia • Stellaria*

CELASTRACEAE

Euonymus or Spindle Tree Family

About 1300 species in 96 genera

RANGE Cosmopolitan but mostly tropical and warm temperate

PLANT FORM Trees, shrubs, and vines, a few perennial herbs; leaves in spirals or opposite, simple, entire, usually with small stipules

FLOWERS Mostly bisexual, radially symmetric, small, usually white or greenish, in cymes, racemes, or solitary; calyx of 2–5 free sepals; corolla usually of 5 free petals but joined into a tube in *Stackhousia*, usually surrounding a well-developed glandular disk; stamens usually 5 (or 10 in 2 whorls), alternating with petals, anthers with longitudinal slits (or transverse in genera not in cultivation); ovary superior, of 2–5 united carpels with as many cells, style 1 with as many stigmas as carpels

FRUIT A capsule, berry, drupe, or winged

Largely a tropical family, few Celastraceae are of horticultural importance. Only species of *Euonymus* are regularly seen in gardens, some grown for their red capsules that split to expose scarlet or orange seeds. Stems of the shrub *E. alatus* (sometimes called burning bush) have striking corky ridges, and the leaves display exceptionally brilliant autumn color. The evergreen *E. fortunei* and *E. japonicus* include shrubs and sprawling ground covers, some with variegated leaves. *Euonymus maakii* is used as a substitute for boxwood in hedges and topiary. The herbaceous genus *Parnassia* (bog star) is occasionally cultivated as an ornamental and has the largest flowers in the family. *Stackhousia*, an Australasian genus of 16 species, some with leaves reduced to scales, is cultivated in Australia, especially *S. monogyna*, known as candles for its attractive, waxy flowers. This genus was treated in the past as a separate family Stackhousiaceae but is convincingly shown by molecular study to be nested within Celastraceae.

The family is recognizable by the simple leaves (with tiny stipules), usually arranged in spirals or sometimes opposite, mostly small, white to greenish or purple flowers with free sepals, usually free petals, and a glandular disk more or less surrounding the superior ovary. Most commonly encountered in gardens, *Euonymus* has reddish capsules that split open to reveal bright red or orange seeds; several species yield useful timber, hence one of its common names, spindle tree. Leaves of the genus *Catha*, known as khat, are chewed in Yemen and elsewhere for their mildly stimulant and euphoric effect. The leaves are harvested from cultivated plantings in Yemen, Ethiopia, and nearby countries.

Selected Genera of Celastraceae
Cassine • Catha • Celastrus • Euonymus • Gymnosporia • Maytenus • Parnassia • Putterlickia • Stackhousia

AQUIFOLIACEAE Although not closely related to Celastraceae, Aquifoliaceae (holly family) are broadly similar in the leathery, spirally arranged leaves and small greenish or white flowers with a superior ovary. The sole genus *Ilex* (holly), with about 600 species of shrubs and trees, has red fruits loosely called berries that are, strictly speaking, drupes. Apart from the distinctive leaves of some *Ilex* species, with the lobes drawn into sharp tips, the flowers are often unisexual, male plants with a rudimentary ovary, female plants with rudimentary stamens. The flowers lack the glandular disk of Celastraceae, and female flowers have a short, stubby style.

CISTACEAE

Rockrose Family

About 175 species in 9 genera

RANGE Northern temperate and subtropics, especially Mediterranean and Middle East

PLANT FORM Shrubs, many trailing, and perennial herbs, often softly hairy, sometimes glandular and producing aromatic ethereal oils; leaves simple, entire, opposite or sometimes in spirals, with or without stipules

FLOWERS Bisexual, radially symmetric, usually in racemes; calyx of 5 sepals, sometimes unequal, green; corolla of 5 (rarely 3) free petals, often crumpled in bud, falling the same day as opening; stamens usually many, free, inserted at base of petals, anthers opening by longitudinal slits; ovary superior, mostly of 3–5(–10) united carpels, 1-celled, usually with many ovules, style 1 or lacking, stigmas 1 (or 3)

FRUIT A capsule with few to many tiny seeds

Although a small family, Cistaceae are disproportionately represented in horticulture, especially in gardens with Mediterranean or warm temperate but relatively dry climates. Several species and cultivars of *Cistus* (rockrose) are grown, some in street plantings. White-flowered *C. ladanifer* produces an aromatic resin used as a scent; and resin from other *Cistus* species may be the myrrh of the Bible. In the wild, *Cistus* species are early colonists after fire. An equally common ornamental in gardens and as mostly gray-leaved trailing plants for rock gardens, *Helianthemum* species and hybrids produce striking displays of white, pink, yellow, or rusty orange flowers. The two genera are separated as follows: *Cistus* has calyx lobes almost equal in size, mostly white to pink petals, and five or ten carpels; *Helianthemum* has a calyx with two narrow outer lobes smaller than the inner, petals often yellow to orange, and three carpels.

Sometimes grown in gardens, the North American *Crocanthemum canadense* is known as frost weed. Species of *Halimium*, sometimes called false sun-rose or simply halimium, are short-lived shrubs ideal for dry situations. The yellow flowers often have a contrasting dark center.

Cistaceae are easily recognized by their simple, opposite leaves, five free petals crumpled in bud and soon falling, many stamens, and a superior ovary. Attractive as the flowers are, the petals fall only hours after opening, and that characteristic combined with the many free stamens render Cistaceae unmistakable.

Selected Genera of Cistaceae
Atlanthemum • Cistus • Crocanthemum • Fumana • Halimium • Helianthemum • Tuberaria

Cistaceae, *Tuberaria guttata*

Cistaceae, *Cistus crispus*

CLETHRACEAE

Clethra Family

About 60 species in 2 genera

RANGE Northern temperate and tropical areas

PLANT FORM Mostly trees, some shrubs; leaves simple, often with serrated margins, in spirals, without stipules

FLOWERS Bisexual, radially symmetric, in racemes or cymes without bracts; calyx of 5 or 6 lobes, green; corolla with 5 or 6 petals free or united basally; stamens twice as many as corolla lobes, free, inserted on petals or corolla tube, anthers opening by pores or short slits; ovary superior, of 3–5 united carpels with 1 to many ovules, stigma terminal, simple or 3- or 4-branched

FRUIT Dry capsules, sometimes splitting open when mature or remaining closed, persisting long after ripening

Clethraceae are a family of just two genera, with *Clethra* contributing several ornamental species, some not well known yet worth attention. The small shrub *C. alnifolia* from eastern North America, sweet pepperbush or summersweet, has racemes of small white flowers. The name pepperbush alludes not to the scent but to the round, peppercorn-like capsules, characteristic of the entire genus. The tree *C. barbinervis* is known for its exfoliating bark, leaving attractive, smooth, cinnamon-colored or pink trunks as well as a profusion of small, white, scented flowers. The lily-of-the-valley tree (*C. arborea*) is native to Madeira.

The family is recognizable by the simple leaves, spirally arranged, small and mostly white flowers with free (or basally united) petals, twice as many stamens as petals, and inverted anthers releasing pollen though pores or short slits, similar to the anthers of Ericaceae (heath or rhododendron family), to which Clethraceae are closely related. The small capsules persist on the plants though the winter and into the next year.

Genera of Clethraceae
Clethra • Purdiaea

Clethraceae, *Clethra barbinervis*

Colchicaceae, *Gloriosa modesta*. Closely allied to Liliaceae, Colchicaceae have a perianth of six petaloid tepals, six stamens, and a superior ovary of three united carpels. The family is separated from Liliaceae by three free styles or, as in *Gloriosa*, the single style with three style branches. Colchicaceae often contain alkaloids, notably colchicine, and lack the raphide crystal inclusions of Liliaceae.

COLCHICACEAE

Colchicum Family

About 210 species in 15 genera

RANGE Most diverse in southern hemisphere in Australia and Africa, also Eurasia and North America

PLANT FORM Herbaceous perennials, mostly with corms (sometimes rhizomes), some stemless, or vines (*Gloriosa*); leaves simple, with parallel veins, sometimes terminating in a tendril, spirally arranged or rarely whorled or in a basal tuft, without stipules

FLOWERS Bisexual, radially symmetric; in racemes, umbels, or solitary; calyx and corolla not differentiated, thus both whorls petal-like and usually similar in size and color, free or united basally or in a tube, tepals 6 in 2 whorls, upright or nodding to pendent; stamens 6, inserted at base of tepals, anthers extrorse; ovary superior, compound, of 3 united carpels with 3 free styles or 1 terminal style with 3 branches, 3-celled

FRUIT Dry capsules, sometimes splitting open when mature and containing brown seeds, or a berry (*Disporum*)

A family of modest horticultural importance, Colchicaceae include the *Crocus*-like genus *Colchicum*, several species of which produce flowers in autumn when leafless, the foliage appearing the following spring and then dying back in summer. Genera occasionally cultivated include the Australian *Burchardia* (milkmaids) and the African *Gloriosa* (for example, *G. modesta*, butter lily or yellow bells) and *Sandersonia*, both trailing and with tendril-tipped leaves. *Sandersonia* (Christmas bells), with striking, urn-shaped, pendent, orange flowers, is cultivated for the florist trade and is desirable for warm temperate gardens.

Several Asian species of *Disporum* and the North American *Uvularia*, especially *U. grandiflora*, are attractive subjects for cool and even cold temperate woodland gardens. Particularly attractive, *D. flavens* has unusual pendent yellow flowers. The North American species included in the past in *Disporum* have been transferred to the genus *Prosartes* of the Asparagaceae (or in some taxonomic systems, the Convallariaceae). The resemblance of *Prosartes* to species of *Disporum* is evidently entirely superficial.

Many Colchicaceae contain poisonous alkaloids, especially colchicine, obtained from corms and seeds of *Colchicum* and still valued medicinally today as a painkiller and for relief from gout. Several other genera of the family also contain colchicine. Among related families, Colchicaceae stand out in lacking the needle-like clusters of calcium oxalate crystals called raphides.

Colchicaceae, *Disporum flavens*

Colchicaceae, *Gloriosa superba*

The family can be recognized among the lily-like families by the three free styles, or if a single style then with three long branches. As in other lilioid (lily-like) families, the anthers are extrorse. The six stamens and superior ovary distinguish the stemless *Colchicum* from *Crocus* (Iridaceae), which has three stamens and an inferior ovary.

Selected Genera of Colchicaceae
Baeometra • *Burchardia* • *Colchicum,* including *Androcymbium* and *Merendera* • *Disporum* • *Gloriosa* • *Ornithoglossum* • *Sandersonia* • *Uvularia* • *Wurmbea*

COMMELINACEAE

Spiderwort Family

About 650 species in 40 genera

RANGE Cosmopolitan, mostly tropical but some cold temperate

PLANT FORM Mostly perennial herbs, many succulent, some annuals and a few vines, with fleshy stems swollen at the nodes and often rooting; leaves simple, in a basal tuft or in spirals, with closed sheath wrapped around stem, parallel veined

FLOWERS Mostly bisexual, radially symmetric or zygomorphic (in tropical genera), 3-merous, ephemeral and lasting only a few hours; calyx of 3 free sepals, usually green, or sometimes petal-like; corolla of 3 free or basally united petals, mostly blue, purple, or white (also pink or yellow), sometimes 1 different from the others; stamens 6 in 2 whorls or 3 with 1 whorl reduced, filaments often bearded with hairs, anthers usually with longitudinal slits, sometimes with expanded connective; ovary superior, of 3 united carpels, 3-celled with 1 to many ovules, style 1 with capitate or lobed stigma

FRUIT Usually a dry capsule

Commelinaceae, *Commelina africana*

Among Commelinaceae in cultivation in temperate horticulture, the most important genus is *Tradescantia* (spider plant), especially the hardy North American *T. virginiana*, of which there are several cultivars with flowers of different colors (blue, violet, white) and leaf variants. Several more species are grown, especially those with trailing stems, not frost hardy but used as ground covers in the tropics and areas of Mediterranean climate. Some are grown as houseplants, especially red-leaved *Rhoeo discolor*. *Gibasis geniculata* and *G. pellucida* (sometimes called Tahitian bridal veil) have dark green, trailing stems and small but attractive white flowers.

The family is recognizable among the monocots by the short-lived flowers, lasting only a few hours, stems swollen at the nodes and rooting from them in trailing species, and leaves forming a sheath wrapped around the stem. The superior ovary bears a single style, and the filaments are often markedly hairy.

Selected Genera of Commelinaceae
Aneilema • Commelina • Cyanotis • Dichorisandra • Gibasis • Palisota • Rhoeo • Tradescantia • Zebrina

Convolvulaceae, *Ipomoea indica*. Vines with spirally arranged, heart-shaped leaves, *Convolvulus* and some species of *Ipomoea*, such as *I. indica*, have a calyx of five sepals subtended by bracts. The five petals are almost entirely united in a funnel-shaped corolla that is furled like an umbrella in bud. The five stamens are appressed to a style that terminates in an enlarged, globular, three-lobed stigma. Bindweeds (*Convolvulus*) have narrow, linear stigmas.

CONVOLVULACEAE

Bindweed or Morning Glory Family

About 1850 species in 55 genera

RANGE Cosmopolitan but mainly tropics and subtropics

PLANT FORM Some shrubs, mainly herbaceous vines, some parasites, a few perennials, usually with milky sap, often with thick rhizomes or tuberous roots; leaves in spirals, entire, lobed, or sometimes pinnately divided (seemingly compound), scale-like in parasitic *Cuscuta*, with small stipules

FLOWERS Usually bisexual, in heads, racemes, or axillary, sometimes solitary, often radially symmetric but a few zygomorphic, usually 5-merous, ephemeral; calyx with sepals at least partly united, green; corolla with petals united and tubular below, often hardly lobed, furled umbrella-like in bud; stamens as many as petals, inserted on lower part of corolla tube, filaments sometimes unequal, anthers with longitudinal slits; ovary superior, of 2 (or 3–5) united carpels, usually with as many cells, with solitary or separate styles, mostly with 2 ovules per cell

FRUIT Usually a dry capsule splitting irregularly or by circular slit, rarely fleshy; seeds often with black, shiny coat

Convolvulaceae are best known as vines, some perennial, others annual, with large flowers each lasting a single day. Flowers are mostly radially symmetric, the corolla tubular below and spreading above but scarcely or only weakly divided into discrete lobes. Most well known are the morning glories, the large flowers opening early in the morning and collapsing early in the afternoon. The common name is loosely applied to species of both *Convolvulus* and *Ipomoea*. The two genera often have very similar flowers and are distinguishable mainly by the style, this divided into linear stigmatic lobes in *Convolvulus* (and the pollen smooth) or not divided in *Ipomoea* (and pollen spiny).

Convolvulus includes the weedy bindweeds, especially *C. arvensis*, invasive and a pest in many places. The genus *Calystegia* is currently considered to belong in *Convolvulus*, and *C. soldanella* has a nearly worldwide range along seashores, the seeds distributed by ocean currents. Unlike most species, *C. cneorum* from southern Europe is a small tufted shrub, hardy in Mediterranean climates, with silvery gray foliage and white flowers. Plants may be treated as annuals in cold climates or overwintered under glass. The blue rock bindweed, *C. sabatius* (also known as *C. mauritanicus*), is a small-flowered, trailing perennial with a woody base, also best treated as an annual in cold climates.

Some species of *Ipomoea* (for example, *I. indica*, purple morning glory), the largest genus of the family, have morning-glory-like flowers almost indistinguishable from

Convolvulaceae, *Convolvulus coelisyriaca*

Convolvulaceae, *Convolvulus cneorum*

Convolvulus, but others such as *I. lobata* (Spanish flag), also known as *Mina lobata*, have narrow, tubular, red and yellow flowers, well worth growing as an annual but not often seen in gardens today. Red-flowered *I. quamoclit* (sometimes placed in the genus *Quamoclit* as *Q. pennata*), the cypress vine, is widely grown in the tropics and as an annual in the temperate zone and is especially attractive to hummingbirds in the Americas. Another striking member of the genus is the moonflower (*I. aculeata* or *Calonyction aculeatum*), which has large white flowers that open at night.

The dried fruits of *Argyraea* and *Merremia* are sold as wood roses in the florist trade. *Dichondra* species are sprawling ground covers sometimes used as lawn plants. Rather more delicate than traditional lawn grasses, *D. micrantha* and *D. repens* (the specific name means creeping) have quite small flowers and kidney-shaped leaves, giving rise to one of its common names, kidney weed. Although difficult to establish, dichondra lawns have the advantage that they need no mowing. *Dichondra micrantha* is native to eastern Asia, and *D. repens* to New Zealand and parts of Australia.

Ipomoea also includes the sweet potato (*I. batatas*), a cultigen native to Central America but dispersed historically to Polynesia, where the swollen, starchy roots are a staple. The sweet potato is now widely cultivated and an important food crop worldwide. The orange-fleshed so-called yams in North American markets are also sweet potatoes; true yams are obtained from some species of *Dioscorea* (Dioscoreaceae). Among numerous cultivars are those grown as trailing ornamentals, favored for their unusual leaf coloring, sometimes dark red to black or pale yellow. Several other *Ipomoea* species are also grown for their starchy tubers. Seeds of some *Ipomoea* species, especially *I. tricolor*, are hallucinogenic; others are purgative.

Convolvulaceae are recognizable by their mostly five-merous, funnel-shaped flowers, furled like an umbrella in bud and usually lasting only a single day, leaves heart shaped, or sometimes lobed, and often vine-like habit, although some are sprawling rather than twining perennials.

Selected Genera of Convolvulaceae
Argyraea • Convolvulus, including *Calystegia • Dichondra • Evolvulus • Falkia • Hildebrandtia • Ipomoea,* including *Quamoclit • Merremia*

CORNACEAE

Dogwood Family

About 90 species in 7 genera

RANGE Mostly northern temperate to sub-Arctic, rare in Africa

PLANT FORM Trees, shrubs, and more or less herbaceous subshrubs, evergreen or deciduous; leaves opposite, in spirals or in two ranks, simple, usually entire, without stipules

FLOWERS Bisexual or sometimes unisexual, radially symmetric, in cymes or clustered in heads, these sometimes surrounded by conspicuous, large, white bracts (1 in *Davidia*); calyx mostly with 4 or 5 (to 10) free, small, scale-like sepals (or forming a tube in male flowers), usually green; corolla with 4 or 5 (to 10) petals, these usually free (or absent in female flowers); stamens as many as petals and alternating with them; ovary inferior, of 2–5(–10) united carpels, with 1 style or separate styles, with 1 (to 5) ovules per cell

FRUIT Varied, a drupe, sometimes fused into a compound berry-like structure, or a large nut

The circumscription of Cornaceae has been in flux for many years, the family sometimes including just two genera or many more, among them *Aucuba* and *Garrya* (now Garryaceae, described below), with *Davidia* and *Nyssa* sometimes recognized as one or two separate families. Most current authorities prefer to include the latter two genera plus their close relatives in Cornaceae, a solution followed here. Cornaceae then include not only *Cornus* (dogwood, osier) but *Davidia* (dove tree), with one species, native to China; *Nyssa* (tupelo), with species in North America and eastern Asia; and several more genera not or hardly known in cultivation.

Several species of *Cornus* are cultivated in temperate gardens, most conspicuously *C. florida* (eastern dogwood), arguably the loveliest of all. *Cornus nuttallii* (Pacific dogwood) is a taller tree with similar flower heads surrounded by large white bracts similar to those of *C. florida*. Summer-blooming *C. kousa* (Chinese, Korean, or kousa dogwood) completes the trio of arborescent and deciduous dogwoods with large bracts surrounding the flower heads. The sub-Arctic, cool temperate, and montane dogwoods *C. suecica* and the very similar *C. canadensis* and *C. unalaschkensis* (dwarf cornels, bunchberry) are dwarf plants, reaching no more than a few inches or centimeters above the ground. All three also have typical dogwood-like flower heads.

Other dogwood species, some trees and some shrubs (osier dogwoods), have flowers more loosely arranged in cymes without the surrounding enlarged bracts. Several are grown for their pagoda-like growth form or their red or yellow-green bark, dramatic in a leafless winter landscape. The late-winter-blooming *Cornus mas* (cornelian cherry) is remarkable for its winter display of bright yellow flowers in tight clusters and lack of conspicuous bracts.

Davidia involucrata, native to interior China, recalls the true dogwoods but has a single large white bract surrounding a crowded head of small flowers. Species of *Nyssa* (tupelo or black gum) are occasionally cultivated for their pyramidal shape and striking autumn color and are used in urban forestry. Several species of Cornaceae are logged for timber, including species of *Nyssa* and *Cornus nuttallii*, tallest of the dogwoods, which can grow into a substantial tree. The cherry-like fruits of *C. mas* are edible and used fresh or dried in the Middle East and eastern Europe, sometimes juiced as a refreshing drink. The fruit is high in vitamin C. The wood of *C. mas* is dense, hard, and durable and was used in antiquity for many purposes, including wheel spikes, bolts, and even weapons.

Cornaceae are recognizable by the small, radially symmetric flowers, usually with scale-like calyx, mostly four or five petals, stamens as many as the petals, and an inferior ovary. Genera with conspicuous white (or pink) bracts surrounding a head-like inflorescence are unmistakably Cornaceae but are actually the exception, not the rule, in the family. Leaves may be opposite, in whorls, or in spirals. The closely related Hydrangeaceae typically have opposite leaves, at least twice as many stamens as petals, and the ovary may be superior or inferior.

Genera of Cornaceae
Alangium • Camptotheca • Cornus • Davidia • Diplopanax • Mastixia • Nyssa

GARRYACEAE We mention the family Garryaceae here as it has traditionally been included in Cornaceae. *Garrya*, a Mesoamerican and western North American genus, is an evergreen shrub or small tree bearing male flowers in attractive pendulous tassels, hence the name silk-tassel tree. The female flowers, concealed by overlapping bracts, are inconspicuous and lack calyx and corolla. *Aucuba*, the other genus of Garryaceae, with three or four species of the Himalaya and eastern Asia, is also evergreen. The cultivar 'Variegata' of *A. japonica* (Japanese laurel) is often grown as a hedge plant and thrives in shady situations.

Cornaceae, *Cornus nuttallii*

Cornaceae, *Cornus controversa*

Cornaceae, *Cornus unalaschkensis*

CRASSULACEAE

Crassula or Stonecrop Family

About 1350 species in 34 genera

RANGE Almost cosmopolitan, especially dry tropics and subtropics, many in southern Africa

PLANT FORM Mostly small shrubs or perennials, a few annuals, all more or less succulent, sometimes almost tree-like with thick fleshy stems, also a few aquatics; leaves simple, often entire, succulent, in spirals, opposite, or in whorls or basal rosettes, without stipules

FLOWERS Bisexual, usually radially symmetric but occasionally zygomorphic, in cymes, racemes, or solitary; calyx mostly with 4 or 5 sepals, usually green; corolla with petals free or partly united, as many as sepals; stamens usually twice as many as petals in 2 whorls or with only 1 whorl alternating with petals; ovary of free carpels as many as petals, superior, each carpel with separate terminal style and containing few to many ovules

FRUIT Usually dry follicles, splitting completely

Crassulaceae, *Crassula columnaris*

Comprising many attractive succulent species, Crassulaceae are well known to gardeners, especially those in semiarid regions. Many species, for example, the jade plant (*Crassula ovata*), are grown indoors. *Sedum* and *Sempervivum* are widely grown in temperate gardens and include many cold-hardy species. The family includes several small, inconspicuous annuals, none of horticultural significance, and several tree-like species, including the southern African genus *Cotyledon*, which can withstand some frost and is a striking addition to gardens with warm temperate to Mediterranean climates.

Crassulaceae can be recognized by their succulent leaves and flowers with free carpels as many as the petals and each with its own style. The family may be confused with the closely related Saxifragaceae, especially those with fleshy leaves and free petals, but Saxifragaceae typically have two at least partly united carpels but also separate styles or at least separate distally.

Selected Genera of Crassulaceae
Adromischus • Aeonium • Bryophyllum • Cotyledon • Crassula • Dudleya • Echeveria • Kalanchoe • Monanthes • Sedum • Sempervivum • Tylecodon • Umbilicus

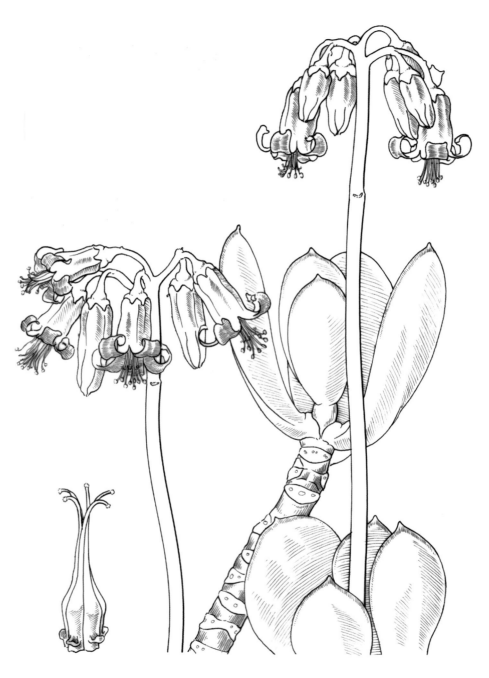

Crassulaceae, *Cotyledon orbicularis*. Like many Crassulaceae, pig's ear is a leaf succulent. The nodding flowers have five free, green sepals, a corolla of five petals united below in a tube, and ten free stamens. The ovary consists of five free carpels, each with its own style.

Crassulaceae, *Crassula perfoliata*

CYPERACEAE

Sedge Family

About 5300 species in 98 genera

RANGE Cosmopolitan but mostly in temperate climates, even subarctic

PLANT FORM Large to small grass-like herbs with rhizomes, mostly perennial but some annuals, a few shrub-like with perennial stems; often growing in wet places, even standing water as floating or submerged aquatics; stems often triangular in cross section (but not invariably); leaves simple, in spirals, often in three ranks, with sheathing base and narrow, strap-like blade with parallel venation, occasionally with short appendage at base of blade (a ligule, also present in true grasses, Poaceae)

FLOWERS Usually unisexual, in spikes or spikelets (or solitary), borne in axils of small bracts, sometimes subtended by chaffy bracts; individual flowers sessile, without differentiated calyx or corolla, tepals scale-like, usually 6 or fewer or lacking; stamens often 3 (or 6) or fewer, anthers with longitudinal slits; ovary superior, usually of 3 united carpels (or 2 or 4), 1-celled with 1 ovule, style terminal, usually with 3 long stigmatic branches

FRUIT An achene, thus with a single seed with a thin covering derived from the outer layer of the ovary

Although a large family, Cyperaceae offer few plants suitable for gardens, although many can be weedy. *Carex*, one of the few genera in wide cultivation, is the largest genus in North America with more than 700 species there alone, and there are almost as many species in eastern Asia. Several species are in cultivation, especially those with leaf pigment mutations, either pale green, yellow, or streaked yellow to white and green, which are particularly useful additions in wet or shady situations. *Uncinia rubra* (red hook sedge) has attractive dark red foliage, thus a particularly attractive addition to low borders.

The most famous of all Cyperaceae is the papyrus plant (*Cyperus papyrus*), depicted in Egyptian hieroglyphics and native along the Nile, where it can grow to 10 feet (over 3 meters) or more tall along riverbanks or in huge floating mats. Historically, strips of the pithy interior of the flowering stems were beaten together to make the paper called papyrus. Species of *Cyperus*, including *C. papyrus*, are grown in water gardens, at pond edges, and even on open ground. The southern African *Ficinia* is sometimes grown for its distinctive, large bracts surrounding the spikelets. The genus *Eriophorum* (Greek, wool bearer) is a pan-northern-temperate, Arctic, and montane genus of herbs with wooly seed heads found in boggy and moorland habitats. The wooly material is derived from the perianth, which grows into wooly threads as the fruits mature and aids their distribution by wind. Known by the common name cotton grass, the wooly threads of different species of *Eriophorum* have been used to stuff pillows and to make paper and cloth.

Cyperaceae, *Carex elata*

One of few foods the family offers is the Chinese water chestnut or simply water chestnut, the corms or tubers of *Eleocharis dulcis*, a tropical Afro-Asian water plant long cultivated in China and eaten raw or cooked. Confusingly, the water caltrop (*Trapa natans*) also provides so-called water chestnuts, which are the seeds of this floating annual of the family Lythraceae.

Cyperaceae are recognizable among the graminoid (grass-like) families by the stems triangular in cross section, sadly not an invariable feature, and the leaf sheaths often with margins united around the stem. In *Carex* and close relatives with unisexual flowers, the female and male flowers are borne on separate parts of the spikelets.

The name bulrush (or bullrush) is a common name applied to large Cyperaceae, including papyrus but also *Typha* (Typhaceae). In common usage, reeds are any of several large grass-like water plants, including some Cyperaceae. Some grasses (Poaceae) and *Typha* are better known as cattails. Rushes are usually *Juncus* (Juncaceae, another grass-like family), but that name has been variously applied to other water plants, including Cyperaceae, The name sedge is specific to Cyperaceae.

Selected Genera of Cyperaceae
Carex • Cyperus • Eleocharis • Eriophorum • Ficinia • Gahnia • Mapania • Mariscus • Scirpus • Tetraria • Uncinia

Ericaceae, *Rhododendron* cultivar of the azalea group. Ericaceae have very diverse flowers but almost always with the petals united basally to almost entirely. Genera share a prominent and frequently persistent style, visible here on the developing fruit, and unusual stamens with the anthers opening by terminal pores.

ERICACEAE

Heath or Rhododendron Family

About 4250 species in 113 genera

RANGE Cosmopolitan but not in arid areas, widespread across northern temperate regions, many in southern Africa, also tropical mountains, especially the Himalaya and South American Andes

PLANT FORM Mostly shrubs, sometimes trees, rarely vines, some growing in waterlogged boggy places, a few parasites, some lacking chlorophyll; leaves simple, in spirals, opposite, or in whorls, often needle-like (so-called ericoid leaves with margins rolled under), without stipules

FLOWERS Usually bisexual, usually radially symmetric, sometimes zygomorphic, usually in bracteate racemes or solitary; calyx mostly of 4 or 5 free (or basally united) sepals, green or petaloid and often persistent, thus visible in fruit; corolla mostly of 4 or 5 petals (as many as sepals), united basally or in a well-developed tube; stamens in 2 whorls, twice as many petals, inserted on base of tube, anthers inverted during development, often splitting by pores or slits and sometimes with appendages (pollen is shed in groups of 4, called tetrads); ovary mostly of 4 or 5 united carpels and as many as 5 locules, superior in subfamily Ericoideae, inferior in Vaccinioideae, each carpel usually with many ovules (rarely 1), style 1, sometimes capitate, often persistent, remaining green as the fruits mature (not in subfamily Vaccinioideae)

FRUIT Usually dry capsules with many minute seeds, or berries (*Vaccinium*), likewise with minute seeds

Several subfamilies of Ericaceae are recognized, most importantly Arbutiodeae (*Arbutus*, *Arctostaphylos*), Ericoideae (*Calluna*, *Daboecia*, *Erica*, *Kalmia*, *Rhododendron*), and Vaccinioideae (*Gaultheria*, *Pieris*, *Oxydendrum*, *Vaccinium*). One of the most important families for horticulture, few gardens are without at least one member of Ericaceae. *Rhododendron*, with about 1000 species, is largely northern temperate, though there are tropical species. The genus now includes *Ledum*, *Menziesia*, and of course *Azalea*, according to current taxonomy. The genera *Pyrola* (wintergreen) and *Monotropa* (Indian pipes) of subfamily Monotropoideae are parasites, the latter lacking chlorophyll. Australian and New Zealand native heaths of the genus *Epacris* have sometimes been regarded as a separate family, Epacridaceae. Most Ericaceae favor well-drained, acidic soils (few tolerate alkaline conditions) and have compact, fibrous root systems.

Rhododendron contributes both wild species and a kaleidoscope of hybrids for horticulture, many hardy but also species of tropical origin belonging to the vireya group. *Erica*, with more than 700 species, is centered in southern Africa; many are grown in warm temperate gardens, but several are cold hardy. Other hardy

Ericaceae, *Erica vestita*

Ericaceae, *Kalmiopsis leachiana*

Ericaceae, *Rhododendron macrophyllum*

Ericaceae, *Rhododendron ciliicalyx*

ornamentals include species of *Andromeda* (bog rose-mary), *Kalmia* (mountain or American laurel, calico bush), *Leucothoe*, *Pieris* (sometimes confusingly called andromeda), and the lesser known but striking shrub *Zenobia*. Several species of *Arctostaphylos* are grown in temperate gardens, some for their blue-gray foliage and remarkable glossy red bark, also a feature of several *Rhododendron* species.

Vaccinium, the genus of blueberries, cloudberries, cranberries, and huckleberries, provides a few garden ornamentals, and several species and hybrids are important in agriculture. Many Ericaceae contain toxic compounds. Leaves of *Kalmia* and some *Rhododendron* species are poisonous to livestock. Excessive eating of *Rhododendron* honey can cause low blood pressure and dizziness, so-called mad honey poisoning. *Andromeda* honey may also cause dizziness and a fall in blood pressure.

Ericaceae can be recognized by their calyx and corolla parts in fours or fives, the calyx persistent and the corolla lobes united at least basally or in a well-developed cup or tube, usually twice as many stamens as corolla lobes, and usually a superior ovary (ovary inferior in the blueberry group, the vaccinioids, *Vaccinium* and relatives). The ovary consists of four or five united carpels with a single style, often persistent and remaining on the developing capsules. The anthers are often specialized, having appendages and frequently opening by pores rather than longitudinal slits. An unusual feature of most Ericaceae is that the pollen is shed in tetrads, that is, four grains united into a single unit. The seeds are always tiny, whether in capsules (then dry and almost dust-like) or in berries. Many Ericaceae have an urn-shaped corolla, another clue to identifying the family.

Selected Genera of Ericaceae
Agapetes • Andromeda • Arbutus • Arctostaphylos • Calluna • Cassiope • Cavendishia • Chimaphila • Daboecia • Enkianthus • Epacris • Erica • Gaultheria • Kalmia • Kalmiopsis • Leucothoe • Oxydendrum • Phylodoce • Pieris • Rhododendron • Vaccinium • Zenobia

Euphorbiaceae, *Euphorbia mauritanica*. Many species of *Euphorbia* are stem succulents with characteristic milky latex and small, deciduous leaves. The unisexual florets are agglomerated in psudanthia (false
flowers) surrounded by fleshy nectar glands that resemble petals. Male florets are reduced to a single
stamen, and the female florets to a three-chambered ovary with a branched style. In *E. mauritanica* the
flowering shoots are terminated by a solitary, nearly sessile pseudanthium containing only male florets
surrounded by a whorl of several smaller, stalked pseudanthia, each containing several male florets and a
solitary female floret.

EUPHORBIACEAE

Euphorbia or Spurge Family

About 5600 species in 214 genera

RANGE Cosmopolitan except arctic and few northern temperate, many in dry tropics

PLANT FORM Trees, shrubs, vines, perennial herbs, and annuals (many weeds), some stem succulents, with milky (or colored) latex; leaves mostly simple, in spirals or whorls, occasionally compound or sometimes leafless, with large or small stipules or stipules lacking; often poisonous, especially the seeds

FLOWERS Radially symmetric, sometimes unisexual, in cymes; without distinction between calyx and corolla, individual tepals small, inconspicuous, or lacking; stamens 5 to many, anthers with longitudinal slits or apical pores; ovary superior, of 3 (or more) united carpels with as many cells as carpels, each with 1 ovule, styles separate or style 1 with much-forked branches

FRUIT Usually a dry capsule splitting at maturity into sections each with single seed, often dispersed explosively, or sometimes fleshy

The flowers of most Euphorbiaceae are difficult to interpret, especially in the genus *Euphorbia* (for example, *E. mauritanica*, yellow milk bush) with some 2000 species worldwide. The individual inflorescences, sometimes called pseudanthia (false flowers), are variously arranged on the stems. Each inflorescence consists of separate male and female flowers, each lacking calyx or corolla. Although largely tropical, the largest genus in the family, *Euphorbia*, comprises several hardy or half-hardy perennials or subshrubs as well as tree-like forms that resemble the saguaro cactus, complete with spines. At least hardy in warm temperate and Mediterranean climates, *E. characias*, *E. cyparissias*, and *E. wulfenii*, along with several more species and hybrids with yellow bracts surrounding flowers, make valuable additions to gardens. The Mexican *E. pulcherrima* (poinsettia), with large red or pale yellow, leafy bracts surrounding the small flowers, is grown as a pot plant sold at Christmas but is hardy in Mediterranean gardens. Many *Euphorbia* species are leafless stem succulents, some reaching tree size, some with spines; smaller species are grown as indoor pot plants. A few other genera are grown for their attractive foliage in the temperate zone, including *Acalypha*.

Many Euphorbiaceae are poisonous, the castor oil plant (*Ricinus communis*) notoriously so. Castor oil, obtained from the seeds, was widely used in the past as a purgative and for lamp oil in antiquity, and it still has many industrial uses as lubricants, in paints, cosmetics, etc. Poisonous alkaloids and proteins in seeds of many other Euphorbiaceae can be fatal. The tubers of the small, tropical tree *Manihot esculenta* are the source of cassava and tapioca, the starch rendered edible by leaching in water. Latex of the tropical tree *Hevea brasiliensis* is harvested for rubber, and the species is

Euphorbiaceae, *Euphorbia globosa*

Euphorbiaceae, *Euphorbia characias*

Euphorbiaceae, *Euphorbia mauretanica*

now cultivated in plantations across the tropics, especially in Malaysia and Thailand.

Selected Genera of Euphorbiaceae
Acalypha • Croton • Euphorbia • Hevea • Jatropha • Manihot • Mercurialis • Ricinus • Xylophylla

PHYLLANTHACEAE Plants in this family were included historically in Euphorbiaceae as subfamily Phyllanthoideae. Phyllanthaceae include few plants for temperate gardens, but the cultivar *Breynia disticha* 'Roseopicta', which has leaves mottled with green, pink, and yellow, is occasionally grown as a ground cover. Like Euphorbiaceae, Phyllanthaceae have milky latex but differ most importantly in having two ovules in each carpel, versus one in Euphorbiaceae.

FABACEAE

Pea Family

About 20,200 species in 741 genera

RANGE Cosmopolitan

PLANT FORM Trees, shrubs, vines, many perennial herbs, some aquatic, and annuals; leaves usually compound, pinnate, bipinnate, or trifoliolate, rarely simple, usually in spirals, with stipules, these often large or sometimes represented by thorns

FLOWERS Usually bisexual, radially symmetric or zygomorphic, in racemes, spikes, or heads; calyx usually of 5 sepals, these sometimes united basally and unequal; corolla of 5 petals, either more or less equal (mimosoids) or somewhat (caesalpinioids) to very (faboids) unequal, then often the upper petal enlarged and upright (called a standard), the lower 2 forming a keel enclosing the stamens and 2 lateral petals lying over the keel as wings; stamens 10 but sometimes some reduced to staminodes (or evidently absent), often with filaments partly united in a sheath around ovary and style (faboids) or many, free, conspicuous, and longer than petals (mimosoids); ovary superior, consisting of 1 carpel, thus 1-celled with few to many ovules in a row and 1 style

FRUIT Usually dry and splitting down two sutures (a legume), sometimes breaking into separate sections, rarely fleshy, containing relatively large hard seeds

One of the largest families of flowering plants, Fabaceae (historical name, Leguminosae) include a substantial number of ornamentals for gardens and many more species used as food and timber. The family is divided into three subfamilies: Caesalpinioideae (caesalpinioids), Mimosoideae (mimosoids), and Faboideae (faboids, also called papilionoids). Although the flowers of each appear very different, all share a similar fruit and fundamental flower structure, and they frequently have compound leaves. Caesalpinioids and mimosoids have flowers radially symmetric to weakly zygomorphic. Mimosoids have many prominent, colored stamens that provide the floral display,

with the small individual flowers crowded in heads or spikes for maximum display. Faboids have markedly zygomorphic flowers, with a large upper petal (the standard) and lateral petals directed forward, and the two lower petals forming a keel enclosing the stamens, ovary, and style. The filaments of faboids, hidden within the keel petals, often form an open tube, with nine filaments joined together and one free and lying over the opening of the filament tube.

Among hardy ornamental perennials and annuals are the faboids *Baptisia*, some *Indigofera* species (many more are tropical), *Lathyrus* (sweet pea), usually vines, and *Thermopsis*. *Lupinus* (the lupin) includes charming, low annuals mostly with blue flowers as well as perennials like the robust hybrids with red, orange, and yellow flowers. Woody faboids include the shrubs *Cytisus*, *Genista*, and *Spartium* (Spanish broom), the vine *Wisteria*, and several trees, most prominent among them red-flowered *Erythrina* (flame or coral trees). Only one or two species of *Erythrina* are hardy, but many more thrive in tropical and subtropical gardens. Other woody faboids include *Cladrastis* (American or Kentucky yellowwood), which has attractive drooping panicles of white or yellow flowers. Although an attractive, fast-growing tree, *Robinia pseudoacacia* (black locust), with honey-scented white flowers, is invasive in Australia, Europe, and southern Africa, spreading by underground sucker shoots.

Caesalpinioids include the woody genus *Cercis*, which has several species, best known of which are the Judas tree (*C. siliquastrum*), a Mediterranean native, and the redbud (*C. canadensis*) of eastern North America, the latter completely hardy. A few *Bauhinia* species are hardy, mostly in Mediterranean or warm temperate gardens, and have distinctive bilobed leaves, recalling those of *Cercis*. Among other caesalpinioids worth horticultural attention is the honey locust (*Gleditsia*), a

Fabaceae subfamily Caesalpinioideae, *Bauhinia galpinii*, Pride of de Kaap. *Bauhinia* is recognizable by its bilobed leaves, with the paired stipules characteristic of the family, and five spreading free petals. The large bean-like pods immediately signify the family. This popular shrub derives its common name from the Kaap River Valley of South Africa, where it is common in the thick bush covering the sides of the valley.

Fabaceae subfamily Faboideae, *Lupinus angustifolius*

Fabaceae subfamily Caesalpinioideae, *Chamaecrista comosa*

very different plant from the black locust; it has inconspicuous flowers but is nevertheless important in urban forestry as a fast-growing and cold-hardy tree. The roasted seeds of Kentucky coffee tree (*Gymnocladus dioicus*) are a substitute for coffee. Hardy, it is also used in landscaping and as street plantings. The unroasted pods and seeds are poisonous.

Ornamental mimosoids include species of *Acacia* from Australia, many with simple, sometimes gray foliage represented by the flattened petiole (called a phyllode); the seedling or juvenile leaves are compound. None is truly cold hardy, but many species thrive in the subtropics and areas of Mediterranean climate. To the distress of scientists and gardeners alike, African and American species until more recently included in *Acacia* are now referred to the genera *Senegalia* and *Vachellia*, some of which are grown in warm temperate gardens. All three genera have flowers in tight heads of yellow or white flowers with the colored filaments providing the floral display. The small tree *Albizia julibrissin* (powder-puff tree) has particularly long stamens with filaments colored white and pink. The dark red-brown foliage of the cultivar called 'Summer Chocolate' is an especially striking plant.

Fabaceae provide many edible seeds (pulses), including beans (mostly *Phaseolus* species), black-eyed peas (*Vigna*), broad beans (*Vicia faba*), chickpeas (*Cicer*), peas (*Pisum*), and lentils (*Lens*). *Medicago* and *Trifolium* are valuable pasture and hay plants, especially alfalfa, also called lucerne (*M. sativa*). Peanuts are the seeds of the Brazilian *Arachis*, the small yellow flowers of which are borne on aerial stems that burrow below ground on elongating stalks as they ripen, hence an alternative common name, ground nuts. Several spices are obtained from Fabaceae, notably tamarind, from the fleshy substance in the pods of the tropical tree *Tamarindus indica*. Licorice is obtained from the roots of the Eurasian *Glycyrrhiza glabra*. Soybeans, the protein-rich seeds of *Glycine max*, are one of the world's most valuable crops. The blue dye indigo is obtained from *Indigofera tinctoria* and some related species.

Ulex europaeus (gorse or furze), which has simple spiny leaves when mature, and *Cytisus scoparius* (most often called Scotch broom) are weedy in many parts of the world, notably in western North America. Unfortunately very resistant to eradication, they cause huge losses in forest production and degradation of pastureland. Some Australian *Acacia* species are invasive in southern Europe and South Africa and likewise invade pastures and native vegetation, causing degradation and loss of native flora.

Selected Genera of Fabaceae subfamily Caesalpinioideae
Bauhinia • Cassia • Cercis • Delonix • Gleditsia • Gynocladus • Senna

Selected Genera of Fabaceae subfamily Faboideae
Arachis • Aspalathus • Astragalus • Baptisia • Cicer • Cladrastis • Clianthus • Coronilla • Crotalaria • Cytisus • Dalea • Dorycnium • Erythrina • Glycine • Glycyrrhiza • Indigofera • Laburnum • Lathyrus • Lens • Lotus • Lupinus • Medicago • Onobrychis • Ononis • Phaseolus • Pisum • Podalyria • Robinia • Sophora • Spartium • Swainsonia • Tamarindus • Thermopsis • Trifolium • Ulex • Vicia • Vigna • Wisteria

Selected Genera of Fabaceae subfamily Mimosoideae
Acacia • Albizia • Calliandra • Leucaena • Mimosa • Parkia • Senegalia • Vachellia

Fabaceae subfamily Faboideae, *Lathyrus odorata*

Fabaceae subfamily Mimosoideae, *Vachellia karoo*

FAGACEAE

Beech or Oak Family

About 970 species in 7 genera

RANGE Mainly northern temperate, also tropical Malesia

PLANT FORM Trees and shrubs, mostly deciduous; leaves simple, in spirals, entire, serrated, or lobed, with stipules soon falling; strongly tannic

FLOWERS Unisexual, usually wind pollinated, inconspicuous; males in pendent catkins or heads, with perianth of (4 to) 6 or more scale-like tepals; stamens 6–12 to many, anthers with longitudinal slits; females several at base of male catkins or in separate axils, enclosed in scaly bracts called involucres; ovary inferior, of 2–7 united carpels with as many cells as carpels, each with 2 ovules, styles as many as carpels

FRUIT A nut with woody or leathery outer layer subtended or enclosed by an accrescent, sometimes spiny involucre (as in chestnuts, *Castanea*) or a cupule (as in the cup in which acorns sit, *Quercus*)

Important forest trees for timber and urban forestry, Fagaceae are well known to most gardeners. The family includes beech (*Fagus*), chestnut (*Castanea*), and oak (*Quercus*). The flowers of most Fagaceae are adapted for wind pollination, hence the catkins of male flowers in deciduous species are produced before the emerging leaves shelter them from the wind. Chestnut flowers alone are pollinated by insects and have scented flowers, albeit unpleasantly so. Widely cultivated for its nuts, *C. sativa*, called Spanish chestnut, yields a valuable timber. *Castanea dentata*, American chestnut, formerly an important timber tree and reputedly producing the finest nuts, is almost extinct due to chestnut blight introduced from Asia in the 19th century. *Castanea mollissima*, Chinese chestnut, is resistant to chestnut blight and has been planted to replace the American chestnut. The western North American golden chinquapin, *Chrysolepis chrysophila* (the generic name means golden scales), is an attractive, small evergreen tree with glossy foliage with golden felt on the undersides. The fruit resembles that of chestnuts and contains edible nuts that taste rather like filberts (*Corylus*).

Both beeches and oaks yield valuable timber, especially for furniture, construction, and in the case of oaks, for shingles, tanning, and casks for aging wine and spirits. The largest genus of Fagaceae, *Quercus* includes some 530 species, ranging from dominant trees in temperate woodland to dwarf trees and shrubs in shrubland in the Mediterranean climate zones of Eurasia and western North America. The Mediterranean *Q. coccifera* or Kermes oak was used in the past for feeding cochineal insects, from which red pigment was obtained for dye. The bark of the cork oak, *Q. suber*, is harvested for cork in 8- to 10-year cycles from trees grown in plantations. Several *Quercus* species are grown as street plantings and as specimen

Fagaceae, *Quercus*, male inflorescences

trees in gardens, among them *Q. ilex* (holm oak), *Q. palustris* (pink oak), *Q. phellos* (willow oak), and *Q. robur* (European or English oak).

Genera of Fagaceae

Castanea • Castanopsis • Chrysolepis • Fagus • Lithocarpus • Quercus • Trigonobalanus

MYRICACEAE Allied to Fagaceae, the Myricaceae (wax myrtle family) include just four genera of which only *Morella* and *Myrica* are sometimes grown as ornamentals in temperate gardens and street plantings. The simple leaves are arranged in spirals, and as in Fagaceae; the inconspicuous flowers are usually unisexual and wind pollinated. The fruits, either small, wax-coated drupes or nuts, are often enclosed in accrescent bracts. The northern-hemisphere *Myrica gale* yields an insect repellent; the leaves are sometimes used to flavor beer, and leaf extracts provide scent for candles. *Morella* includes *M. californica*, the bayberry, and *M. cerifera*, the wax myrtle of eastern North America; wax for candles is obtained from several species.

NOTHOFAGACEAE The southern beeches (*Nothofagus*) of Australasia and southern South America were long included in Fagaceae, but DNA studies show them to be better regarded as a separate family. A genus of evergreen and deciduous trees, *Nothofagus* differs from Fagaceae in details of the inflorescence, pollen, and seeds. Species yield timber for furniture, flooring, and construction.

GENTIANACEAE

Gentian Family

About 1850 species in 95 genera

RANGE Almost cosmopolitan but mostly temperate and subtropical, and tropical mountains

PLANT FORM A few trees, some shrubs, and many perennial herbs, some acaulescent; leaves opposite, rarely in whorls or spirals, simple, entire, smooth, without stipules

FLOWERS Perfect, radially symmetric, in cymes or solitary, rarely in racemes; calyx of 4 or 5 united sepals forming a bell- or funnel-shaped tube; corolla with 4 or 5 united petals also forming a bell- or funnel-like tube; stamens as many as sepals and corolla lobes, alternating with corolla lobes, inserted on corolla tube, rarely some stamens reduced or lacking, anthers with longitudinal slits (rarely pores); ovary superior, compound, of 2 united carpels, 1-celled with many ovules, style 1 with 2-lobed stigma

FRUIT Usually a capsule, sometimes a berry, with small to tiny seeds, often needing specific fungal associations to germinate successfully

Although a relatively large family, Gentianaceae contribute only modestly to horticulture. The genus *Gentiana* has more than 360 species, several of which are grown in gardens and desirable for their large, usually deep blue flowers. Among these is the striking, large-flowered, alpine *G. acaulis*, ideal for rock gardens. The willow gentian, *G. ascelpiadacea*, has willowy stems of dark blue or white flowers, and unusual in the genus, *G. lutea* has yellow flowers.

The lisianthus of floristry is a species of *Eustoma*, an attractive garden annual with blue-gray foliage. The genus *Lisianthus* is also a member of Gentianaceae and includes woody shrubs and perennials. The Mexican *L. nigrescens*, which has remarkable black flowers, is occasionally grown in gardens but needs protection in winter. Species of the large African genus *Chironia* and its relative *Orphium* (for example, *O. frutescens*, sea rose) have brilliant pink flowers, too seldom seen in gardens.

The arborescent genera of the family are hardly known, but the African and Madagascan *Anthocleista* (included in the past in the related family Loganiaceae, described below) is a particularly attractive tree, having huge glossy leaves clustered at the tops of particularly long branches. Some woody Gentianaceae yield useful timber; others have been used in traditional medicine. Extracts of gentian have been used to treat digestive problems, fever, hypertension, muscle spasms, parasitic worms, wounds, cancer, sinusitis, and even malaria. Unfortunately, there is little or no scientific evidence to support the reputed medicinal value.

Gentianaceae, *Orphium frutescens*. Gentianaceae are recognizable by their simple, opposite leaves without stipules, a tubular or bell-shaped calyx of four or five united sepals, and a funnel-shaped corolla with the same number of lobes united for part of their length. Four or five stamens are inserted on the corolla tube, and the superior ovary is concealed within the calyx. The stigma is typically two-lobed, but in *Orphium* it is simple and pad-like. Also unusual for the family are the finely felted hairs covering the vegetative parts and the curiously twisted anthers that release pollen through apical pores.

Gentianaceae, *Gentiana acaulis*

Gentianaceae, *Gentiana carinata*

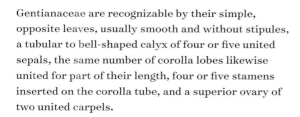

Gentianaceae are recognizable by their simple, opposite leaves, usually smooth and without stipules, a tubular to bell-shaped calyx of four or five united sepals, the same number of corolla lobes likewise united for part of their length, four or five stamens inserted on the corolla tube, and a superior ovary of two united carpels.

Selected Genera of Gentianaceae
Canscora • Centaurium • Chironia • Erythraea • Eustoma • Exacum • Fagraea • Gentiana • Halenia • Lisianthus • Orphium • Potalia • Sebaea • Swertia

LOGANIACEAE Closely allied to Gentianaceae, the Loganiaceae are distinguishable from that family only with difficulty, sometimes by the presence of stipules at the base of the leaves. The family contributes little to horticulture, the only ornamental of significance being the eastern North American *Spigelia marilandica*, a perennial woodland herb with large red flowers. The largely tropical tree *Strychnos*, the source of the deadly poison strychnine, is a member of Loganiaceae. *Buddleja*, placed in older classifications in Loganiaceae, has now been shown to belong in Scrophulariaceae.

Geraniaceae, *Pelargonium ×hortorum*. The umbrella-like leaves with petiole toward the center, and conspicuous stipules, are characteristic of Geraniaceae. Unlike the radially symmetric flowers of most genera, including the true geraniums, the flowers of *Pelargonium* are zygomorphic, with two upper petals and three lower alternating with the green sepals. The ten stamens are joined at the base, but only some are fertile. The fruit has the stiff beak typical of all Geraniaceae and when ripe splits into five separate units at first held together at the tips, each bearing a solitary seed.

GERANIACEAE

Cranesbill Family

About 809 species in 5 genera

RANGE Almost cosmopolitan but mostly temperate zone and tropical mountains

PLANT FORM Shrubs, perennials, and some annuals, sometimes geophytes with tubers, some acaulescent; leaves usually in spirals (rarely opposite) or a basal tuft, usually simple, often toothed or pinnately or palmately lobed to highly dissected, rarely compound, usually with stipules; frequently aromatic with secretions from glandular hairs

FLOWERS Perfect, radially symmetric or zygomorphic, in cymes; calyx of 4 or 5 sepals; corolla with 4 or 5 free petals, often soon falling, rarely petals lacking; stamens 5, 10, or 15 in 1, 2, or 3 whorls, respectively, or some stamens reduced and without anthers, anthers with longitudinal slits; ovary compound, of 5 (or 2 or 3) united carpels, superior, 2-, 3-, or 5-celled, usually 2 ovules per cell, with 1 style and separate stigmas, these persisting and elongating in a prominent column or beak

FRUIT Distinctive capsules, splitting into five one-seeded sections (mericarps), sometimes explosively, to release seeds

Geraniaceae are a family of just five genera, four in cultivation. Widely grown in temperate gardens, *Geranium* has radially symmetric flowers with ten stamens and palmately lobed to dissected leaves. Several species are currently available from nurseries, but many more have horticultural merit. Also with radially symmetric flowers, *Monsonia* has 15 stamens; species include those in section *Sarcocaulon* (no longer regarded as a separate genus) with thick, fleshy stems and spines derived from the leaf petioles. *Erodium*, a largely Eurasian genus, has only five stamens and radially symmetric or weakly zygomorphic flowers with two petals having contrasting markings. Species of *Erodium* are valued as fodder and pasture plants and have been introduced in farmland in many parts of the world; some are now invasive.

Pelargonium, a genus centered in southern African and most diverse there, also occurs in Asia, Australia, New Zealand, and on some Atlantic islands. Flowers are always zygomorphic with the two upper petals largest and with contrasting markings. Although the flowers bear ten stamens, only two to seven of them bear anthers; the remaining ones are sterile. The upper sepal has a nectar-bearing spur fused to the flower stalk, this then seeming to be hollow. The spur contains nectar, important for potential pollinators visiting flowers. The so-called geraniums of the florist and nursery trade are cultivars and hybrids of *Pelargonium* (for example, *P. ×hortorum*, zonal geranium). Few pelargoniums are cold hardy and are best grown as annuals or overwintered in the greenhouse.

Geraniaceae, *Monsonia speciosa*

Geraniaceae, *Pelargonium cucullatum*

Geraniaceae, *Geranium incanum*

Geraniaceae are readily recognizable by their unusual beaked capsules but also by the four- or five-merous flowers with free sepals, deciduous petals, and superior ovary. Although the fruits of all genera are remarkably similar, common names reflect generic differences, thus cranesbill is used for *Geranium*, heron's bill for *Erodium*, and storksbill for *Pelargonium*.

Genera of Geraniaceae

Erodium • Geranium • Hypseocharis • Monsonia, including *Sarcocaulon • Pelargonium*

MELIANTHACEAE Allied to Geraniaceae, Melianthaceae include some 13 species of trees and shrubs in three genera, all native to sub-Saharan Africa. The simple or pinnately compound leaves are arranged in spirals and in *Melianthus* have prominent stipules. *Melianthus major*, which has large, handsome gray leaves, is half-hardy and useful in Mediterranean and warm temperate gardens. It will survive some freezing and can be treated as a perennial in marginal situations and in coastal northwestern North America. The southern African genus *Greyia* includes three species of woody shrubs with attractive red flowers displayed in compact racemes and, like *M. major*, can be grown in warmer climates and will tolerate some frost.

GOODENIACEAE

Goodenia Family

About 410 species in 10 genera

RANGE Mainly Australia, also Chile, New Zealand, and the Atlantic, Indian, and Pacific ocean coasts

PLANT FORM Mainly perennial herbs, shrubs; leaves mostly in spirals, often with axillary tuft of hairs

FLOWERS Bisexual, radially symmetric or more often zygomorphic, in cymes, racemes, umbels, heads, or solitary; calyx of 5 sepals, 3 reduced or vestigial, sometimes united basally; corolla of 5 petals united below into a tube more deeply split on 1 side, then fan-like with 2 unequal lips; stamens 5, opposite sepals, anthers free or joined together, forming a cylinder around style, with longitudinal slits; ovary superior, half-inferior, or inferior, of 2 united carpels, mostly 2-celled (rarely 4-celled or apparently 1-celled), style 1, rarely 2- or 3-branched, stigma surrounded by a cup-like structure

FRUIT A drupe, a capsule rarely separating into one-seeded woody sections, or a nut

Goodeniaceae, *Scaevola aemula*

Goodeniaceae, *Scaevola plumieri*

The almost exclusively southern-hemisphere Goodeniaceae are largely Australian and contribute modestly to temperate horticulture, but several species are grown in gardens with a Mediterranean or subtropical climate. The corolla, always with petals partly united, is unusual in being unequally divided on one side, sometimes to the base, resulting in a fan-like flower. This condition recalls that in *Lobelia* and its close relatives of the Campanulaceae, which like Goodeniaceae have the anthers joined together. As a result, Goodeniaceae have long been considered to be allied to Campanulaceae. Although they lack the milky latex of Campanulaceae, molecular studies confirm this relationship.

Species of *Leschenaultia* are particularly attractive. The hop goodenia (*Goodenia ovata*) is cultivated in Australia as a useful ground cover for rock gardens and open slopes. A compact, low-growing, evergreen shrublet, it produces clusters of bright yellow flowers shaped like little fans. *Brunonia australis* (blue pincushion or Australian cornflower) is an herbaceous perennial with rounded heads of blue flowers resembling those of the cornflower (*Centaurea cyanus*) of the daisy family (Asteraceae).

The red- or blue-flowered *Scaevola aemula* is sometimes cultivated as an annual and is an especially attractive plant for hanging baskets with its trailing stems. Two other species of the genus, *S. plumieri* and *S. sericea*, grow on beaches on the Atlantic, Indian, and Pacific oceans. Their flowers have the fan shape typical of the family. Their seeds float in salt water but germinate only in freshwater.

Selected Genera of Goodeniaceae
Anthotium • Brunonia • Dampiera • Goodenia • Leschenaultia • Scaevola

HAEMODORACEAE

Bloodroot Family

About 86 species in 8 genera

RANGE Mainly Australia and southern Africa, also tropical and North America

PLANT FORM Perennial herbs with rhizomes or tubers, often with red pigment in roots and rhizomes; leaves mostly basal, in two ranks, strap- or sword-like, some with pleated blades, parallel veined, without stipules

FLOWERS Perfect, radially symmetric or zygomorphic, on stems with reduced leaves, in panicles, cymes, or racemes; without differentiated calyx and corolla, thus perianth with 2 whorls of 3 tepals each, free or united in a straight or curved tube; stamens 3 or 6, free or inserted on perianth tube, anthers with longitudinal slits; ovary superior or inferior, of 3 united carpels, 3-celled, with 1 to many ovules, style 1

FRUIT Usually a capsule or nut-like (*Phlebocarya*)

Best known among Haemodoraceae are the kangaroo paws of Australia, *Anigozanthos* species, several of which are cultivated for their unusually colored and sometimes spectacular flowers. None is truly hardy to our knowledge, but they thrive in gardens with Mediterranean and warm temperate climates, especially in their native Australia. Some species of the southern African *Wachendorfia* are occasionally seen in gardens, and at least the marsh and streamside plant *W. thysiflora* is somewhat hardy. Another species with similar, large, pale to deep yellow flowers, *W. paniculata*, also deserves horticultural attention.

Genera of Haemodoraceae

Anigozanthos • Barberetta • Conostylis • Dilatris • Haemodorum • Lachnanthes • Phlebocarya • Wachendorfia

Haemodoraceae, *Dilatris viscosa*

Hamamelidaceae, *Trichocladus grandiflorus*. Comprising trees and shrubs, the Hamamelidaceae include many genera important in temperate horticulture. *Trichocladus* is an exception, for the species are not hardy. The simple, spirally arraned leaves lack stipules, and the unusually large flowers have a five-parted corolla. Anthers dehisce by flaps (valves), and the ovary has two separate styles.

HAMAMELIDACEAE

Witch Hazel Family

About 100 species in 28 genera

RANGE Northern temperate Eurasia Asia and North America, also tropical Asia and Africa

PLANT FORM Large and small deciduous trees and shrubs, branches usually with velvety or felted hairs; leaves in spirals, simple, often toothed, pinnately veined or often with multiple main veins from the base, thus palmately veined, with stipules; several species bloom in late fall, winter, or early spring, thus flowering in cold weather

FLOWERS Perfect, radially symmetric, borne at branch tips or on short shoots in small head-like clusters or drooping spikes; calyx with sepals partly united, 4- or 5-lobed, green or red; corolla with 4 or 5 free petals; stamens as many as petals, sometimes twice as many then some scale-like, inserted on floral disk; ovary superior, compound, of 2 closely united carpels, 2-celled, each with a short style

FRUIT A woody or horny capsule, splitting from the apex and expelling one or few seeds by force; dry capsules often persisting a year or more, sometimes several clustered together in a cone-like ball (*Liquidambar*)

No temperate garden should be without one or more species of Hamamelidaceae for their late winter or early spring flowers produced when little else is in bloom. *Hamamelis mollis* (Chinese witch hazel) and *H. virginiana* (American or common witch hazel) are fine garden subjects, as is the eastern North American *Fothergilla*, with just two species. *Parottia persica* (Persian ironwood), the only species of the genus, is an attractive tree with mottled bark and exceptional fall color; even the young foliage, produced in late spring, has a red tinge. Its inconspicuous flowers are produced in late winter and noticeable, if at all, for the bright red stamens. The display in *Fothergilla* flowers is provided by enlarged, fleshy filaments; the calyx and corolla are inconspicuous.

Also early spring blooming, *Corylopsis* (winter hazel) includes two notable flowering shrubs, *C. pauciflora* and *C. spicata*, which have drooping spikes of yellow blooms borne on leafless branches. *Rhodoleia* is seldom seen in gardens, but at least *R. forrestii* is hardy and has attractive, some say striking, pink flowers. *Loropetalum chinense* is sometimes cultivated, especially the purple-leaved and pink-flowered cultivars (the wild form has white flowers and green foliage). It is not truly hardy, but in mild coastal climates and areas with a Mediterranean climate it is a particularly notable shrub, sometimes sold under the name Chinese fringe tree (easily confused with *Chionanthus retusus*, the Chinese fringe tree of the olive family, Oleaceae). The flowers, with narrow, strap-like, somewhat crumpled petals recall those of the witch hazels. *Trichocladus grandiflorus* (African or green witch hazel) is a particularly handsome tree with coppery new foliage and clusters of

Hamamelidaceae, *Hamamelis mollis*

Hamamelidaceae, *Fothergilla major*

Hamamelidaceae, *Corylopsis spicata*

pink flowers that resemble those of the true witch hazels. Hardly known in cultivation, it would make a valuable addition to gardens with mild, temperate, or Mediterranean climates.

Altingia and *Liquidambar* are valuable timber trees, providing hardwood. The eastern North American *L. styraciflua* (American sweet gum or simply sweet gum, or alligator wood) is widely used in street plantings and, like many Hamamelidaceae, has dramatic autumn color.

Hamamelidaceae are recognizable by their simple, often toothed, spirally arranged leaves, flowers in tight clusters or axillary, with four or five free petals, an ovary with two styles, and horny or woody capsules opening at the apex and usually persisting on the leafless stems. The petals of many genera are narrow and distinctively crumpled.

Selected Genera of Hamamelidaceae
Altingia • Corylopsis • Fothergilla • Hamamelis • Liquidambar • Loropetalum • Parrotia • Rhodoleia • Sinowilsonia • Sycopsis • Trichocladus

ITEACEAE Related to Hamamelidaceae (but in the past included in Saxifragaceae), Iteaceae comprise two genera of which only *Itea* is cultivated in temperate gardens. A shrub or small tree, *Itea* has simple, spirally arranged leaves with minutely to prominently toothed margins, five-merous flowers often in long, drooping racemes, and a superior to almost inferior ovary. The eastern and southern North American *I. virginica* (Virginia sweetspire) is a sprawling deciduous shrub with relatively large (for the genus) fragrant, white flowers. The leathery leaves turn red to purple in autumn and persist well into winter. Other species, including *I. ilicifolia* (holly-leaved sweetspire), are evergreen, and although they have smaller flowers, the long trailing racemes are remarkably attractive, recalling the drooping catkins of *Garrya*, the silver silk-tassel tree (Garryaceae are described following Cornaceae).

HIPPOCASTANACEAE

Horse Chestnut Family

About 20 species in 1–3 genera

RANGE Northern hemisphere, mainly North America and eastern Asia

PLANT FORM Trees and shrubs, mostly deciduous; leaves opposite, palmately compound, with three to eight leaflets with toothed margins, without stipules

FLOWERS Perfect, in terminal panicles, zygomorphic; calyx of 4 or 5 free sepals united basally; corolla of 4 or 5 petals, unequal, clawed, often brightly colored; stamens 5–9, arched downward, filaments unequal, anthers with longitudinal slits; ovary superior, of 3 united carpels, 3-celled, each with 2 ovules, with 1 style

FRUIT A leathery capsule, surface spiny, warty, or smooth, with large shiny seeds (resembling chestnuts, *Castanea*)

Hippocastanaceae, *Aesculus ×carnea*

Hippocastanaceae are an important family for urban forestry. Species of *Aesculus* (buckeyes or horse chestnuts), mostly *A. hippocastanum*, are grown as street trees but are usually too large for small gardens. More suitable for smaller gardens are shrubby species such as *A. parviflora* and the red-flowered *A. pavia*. The California native *A. californica* is a small tree truly hardy in only in Mediterranean and temperate, maritime climates but surviving light freezes.

Hippocastanaceae are recognizable by their opposite, palmately compound leaves, zygomorphic flowers with five to nine stamens with unequal filaments, and a superior, three-celled ovary with a single style. The leathery, thick-walled capsules, sometimes with a spiny surface and large shiny seeds, are distinctive. The seeds resemble those of chestnuts (*Castanea* in the Fagaceae), hence the common name.

Genera of Hippocastanaceae
Aesculus • *Billia,* sometimes included in *Aesculus* • *Handeliodendron,* sometimes included in *Aesculus*

SAPINDACEAE Hippocastanaceae have been included in Sapindaceae (soapberry family) by some authorities; others prefer to recognize Hippocastanaceae as well as the maple family, Aceraceae (discussed alphabetically in the Families A–Z), and Xanthocerataceae (including only the large-flowered *Xanthoceras sorbifolium*) rather than include them in Sapindaceae. We find it most useful to keep these three families separate, but readers should be aware of the alternative treatment that many find acceptable. Although closely related, Sapindaceae (about 1335 species in 130 genera) differ in usually having small flowers and leaves compound and paripinnate in spirals (versus opposite and simple or compound, and palmate or imparipinnate, in some *Acer* species and *Xanthoceras*) and different fruits. Genera of Sapindaceae in cultivation include the shrub or small tree *Dodonaea* (especially red-leaved cultivars of *D. viscosa*) and the tree *Koelreuteria* (the deciduous *K. paniculata* is widely grown as a street tree and is cold hardy). Most Sapindaceae are evergreen, tropical trees and shrubs, but *Cardiospermum* is a vine. The tropical Asian *Litchi* provides the delicious fruit, the litchi or lichee, the fleshy part an aril surrounding the single large seed.

HYACINTHACEAE

Hyacinth Family

About 800 species in 29 genera

RANGE Mostly southern Africa and western Eurasia, also tropical African mountains, one genus in South America

PLANT FORM Small to large perennials with fleshy bulbs; leaves several or few, crowded basally, simple, usually strap-like; flowering stalks always without leaves or bracts; several species bloom in late winter and early spring, thus flowering in cold weather

FLOWERS Perfect, radially symmetric or zygomorphic, borne in bracteate racemes; perianth not differentiated into calyx and corolla, both whorls petal-like, with 6 tepals free or partly united and then 6-lobed; stamens as many as tepals, in 2 whorls, sometimes 1 whorl shorter or sterile, thus staminodes, filaments free or united; ovary superior, compound, of 3 united carpels, 3-celled, with 1 terminal style

FRUIT A dry, thin-walled capsule, splitting from the apex; seeds black, shiny or matte

With four subfamilies, two of them important in horticulture, Hyacinthaceae include the Hyacinthoideae with scilla-like plants from Eurasia and Africa, notably the southern Africa genus *Lachenalia*, and Ornithogaloideae, mainly from Africa but *Ornithogalum* also in Eurasia. The genus *Scilla* has been dismembered, and several erstwhile *Scilla* species are now regarded as different genera. The well-known and misnamed *S. peruviana* (from Morocco) is now *Oncostemma peruviana*, and the African *S. natalensis* is *Merwilla plumbea*. In similar vein, the bluebells, *S. hispanica* and its close allies, are referred to the genus *Hyacinthoides*. Remarkably similar to *H. hispanica*, the British bluebell is *H. non-scripta*, also segregated as the genus *Endymion*. Time will tell which of these changes will be found most acceptable.

Important in horticulture are *Hyacinthus orientalis* (the hyacinth), late winter flowering *Chionodoxa* (glory of the snow), several species of *Muscari* (grape hyacinth), *Hyacinthoides*, and the true scillas or squills, *Scilla*. *Lachenalia* is also becoming important in areas of mild climate with cool but not freezing conditions. Most species of Hyacinthaceae are poisonous to mammals; the bulbs and foliage contain cardiac glycosides and various alkaloids.

Hyacinthaceae are recognizable by their fleshy, onion-like bulb, completely leafless flowering stem, bracteate racemes of flowers consisting of two petaloid whorls of tepals, six stamens, and a superior ovary. Some systems of family classification include Hyacinthaceae in Asparagaceae as subfamily Hyacinthoideae.

Selected Genera of Hyacinthaceae
Albuca • Alrawia • Bellevalia • Chionodoxa • Drimia • Endymion • Eucomis • Hyacinthoides • Hyacinthus • Lachenalia • Massonia • Merwilla • Muscari • Oncostemma • Ornithogalum • Scilla • Veltheimia

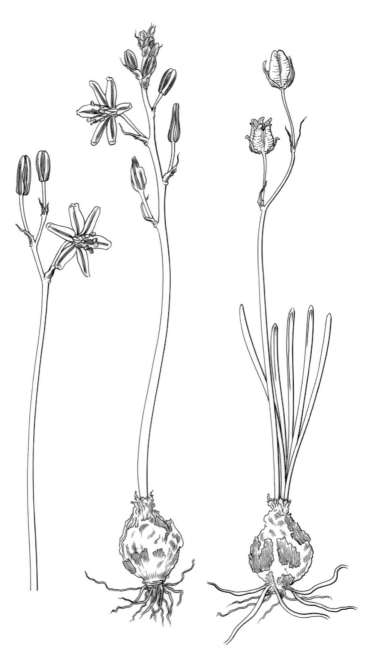

Hyacinthaceae, *Albuca albucoides*. Although sometimes included in an expanded Asparagaceae, Hyacinthaceae are distinguishable by their onion-like bulb, formed by the swollen bases of the leaves, and leafless raceme of more or less regular flowers with six petal-like corolla segments, six stamens, and a three-lobed, superior ovary with a simple style.

Hyacinthaceae, *Ornithogalum narbonense*

Hyacinthaceae, *Merwilla plumbea*

Hyacinthaceae, *Ornithogalum rupestre*

Hyacinthaceae, *Veltheimia capensis*

Hydrangeaceae, *Philadelphus* cultivar. This mock orange is a shrub with toothed, opposite leaves with the petioles distinctively joined across the nodes. The radially symmetric flowers, borne in small paniculate cymes, have four sepals and four free petals that along with the numerous stamens are deciduous. The ovary is inferior, with a single style with four slender branches.

HYDRANGEACEAE

Hydrangea Family

About 185 species in 12 genera

RANGE Himalaya to Japan, North America

PLANT FORM Small trees, shrubs, and woody vines, both evergreen and deciduous, and a few perennial herbs; leaves simple, usually opposite, sometimes in whorls, often toothed, entire or lobed, without stipules

FLOWERS Perfect, radially symmetric, or outer flowers of inflorescence sterile and with enlarged, unequal calyx lobes, in paniculate or corymbose cymes; calyx of 4 or 5 (to many) free or basally united sepals; corolla of 4 or 5 (to many) petals free or basally united; stamens 1 to several, many in *Carpenteria* and *Deinanthe;* ovary superior or partly to fully inferior, of 2–5(–12) united carpels, several- to 1-celled, usually with multiple styles

FRUIT A capsule or berry

Hydrangeaceae are a family of several ornamental shrubs and small trees, the best known of which is *Hydrangea.* The rounded to flat-topped, many-flowered inflorescences have the outer flowers sterile and with large white or colored calyx lobes. In many cultivars of *H. macrophylla*, the so-called mopheads, the entire inflorescence may consist of flowers with conspicuous calyx lobes. *Hydrangea paniculata* and *H. tomentosa* may grow to moderate-sized trees. Several species of *Deutzia* and *Philadelphus* also have horticultural merit; both have moderate-sized white (rarely pink) flowers, some species being particularly fragrant, especially *P. coronarius* (mock orange).

The California native *Carpenteria* has large white flowers notable for the remarkable number of stamens. Rare in cultivation, the herbaceous perennial *Deinanthe* has attractive pale blue flowers, very like those of *Carpenteria* in their numerous stamens, with a nodding and smaller corolla. *Deinanthe bifida* has perhaps the oddest leaves of flowering plants, deeply forked into two lobes in the upper half.

Occasionally seen in temperate gardens and reasonably hardy, *Dichroa febrifuga* has rather modest pink or blue flowers and is grown for its large, brilliant blue berries in autumn and winter as well as its flowers. As its species name indicates, it has been used traditionally to reduce fever and, especially during the Second World War, as an antimalarial and quinine substitute. A last ornamental, the eastern Asian *Kirengeshoma*, with just two species, is another herbaceous genus, and unusual for Hydrangeaceae, it has large flowers of a rich yellow, unfortunately soon dropping its petals.

Hydrangeaceae, *Carpenteria californica*

Hydrangeaceae, *Deutzia gracilis*

The family is readily recognizable by its sterile outer flowers of the inflorescence, with enlarged calyx lobes when these are present, but if not, simple, nearly always opposite leaves, flowers with many stamens, and often a partly to fully inferior ovary, usually with multiple styles. Somewhat distinctive, the petioles are slightly thickened basally so that they meet or are joined by a narrow line across the stem,

Genera of Hydrangeaceae

Broussaisia • Cardiandra • Carpenteria • Deinanthe • Deutzia • Dichroa • Fendlera • Hydrangea • Jamesia • Kirengeshoma • Philadelphus • Whipplea

Iridaceae, *Freesia refracta*. The African genus *Freesia* is known for its attractive flowers, often with a fine fragrance. Characteristic of Iridaceae are the fan of leaves oriented edgewise to the stem, three stamens inserted opposite the outer tepals, and inferior ovary with a single style divided at the tip.

IRIDACEAE

Iris Family

About 2500 species in 65 genera

RANGE Almost cosmopolitan, especially areas of Mediterranean climate, tropical highlands and mountains, especially southern Africa

PLANT FORM Perennials with rhizomes or corms, some evergreen, a few woody shrubs, and a handful of annuals (*Sisyrinchium*); leaves simple, often unifacial, in two ranks and oriented in a single plane (*Iris*-like) or bifacial, with or without main vein, plane or pleated, lacking stipules

FLOWERS Perfect, radially symmetric or zygomorphic, either arranged in clusters enclosed by a pair of large leafy bracts called spathes, emerging one by one from spathes, or in spikes and the flowers then sessile; calyx and corolla not differentiated, thus both whorls petal-like and usually similar in size and color, free or united basally or in a narrow or wide tube, tepals 6 in 2 whorls, upright or nodding to pendent; stamens 3 (or 2), opposite outer tepals, inserted on perianth tube or base of tepals, anthers extrorse; ovary inferior (superior in Tasmanian *Isophysis*), compound, of 3 united carpels, 3-celled, with 1 terminal style, apically usually divided into discrete branches, these sometimes petal-like (for example, *Iris*), with stigmatic surfaces along style branches or as discrete lobes

FRUIT A dry capsule, often with round or flattened seeds, occasionally with fleshy appendages or fleshy seed coat

Iridaceae include many genera of horticultural importance, notably *Iris* and *Gladiolus*, each with about 270 species, and the southern African *Freesia* (for example, *F. refracta*), *Ixia*, and *Watsonia*. Cultivation of *Freesia* and *Gladiolus* hybrids for the cut-flower market is a major industry. The Eurasian genus *Crocus* is widely cultivated in cool and cold temperate areas for its early spring display of flowers, especially *C. vernus* and hybrids derived from that species, but many more species are also grown. The largely African *Dietes*, which is evergreen, is widely cultivated in parks and street plantings, especially *D. bicolor* and *D. grandiflora*. The North and South American genus *Sisyrinchium* is also becoming a popular garden plant, notably the Chilean *S. striatum*, and the so-called blue-eyed grasses, derived from cold-hardy North American wild species. With particularly striking flowers, *Tigridia pavonia* is grown for its large, brilliant red or yellow and even white, spotted blooms. Evergreen New Zealand and South American species of *Libertia* also find a home in many gardens, sometimes for their colored foliage, as well as their fleeting white flowers.

Too many *Iris* species are cultivated to mention all, but bearded irises (mostly hybrids of several species) are grown or persist in many gardens as do cultivars of *I. siberica*, the Siberian iris. Pacific coast irises include wild species, especially

Iridaceae, *Iris missouriensis*

Iridaceae, *Iris iberica*

Iridaceae, *Moraea tulbaghensis*

Iridaceae, *Freesia occidentalis*

I. douglasiana, and a range of hybrids derived from crosses between wild species of California and Oregon. The eastern Asian *I. domestica* (in the past known as *Belamcanda chinensis*), the blackberry lily, is another attractive species for gardens and has become weedy in parts of North America.

The family has contributed only modesty to medicine with the notable exception of *Crocus*, the source of saffron, obtained from the stigmas and style branches of *C. sativus*, and historically from some wild species, one of more of which gave rise to the cultivar, a sterile triploid. Saffron is one of the world's most expensive spices due to the labor involved in extracting the styles and stigmas from the flowers, which are harvested as soon as they open. Historically, saffron was also used for medicinal properties. Corms of some *Crocus*, *Gladiolus*, and *Moraea* species are used locally as food for their starchy content, but some moraeas are toxic to stock and even to humans.

Iridaceae are readily recognizable by their six-parted perianth, three stamens opposite the outer tepals, inferior ovary, and often unifacial leaves arranged in a fan. The Australian *Diplarrhena* stands out in having only two stamens. Rarely seen in cultivation, the Tasmanian *Isophysis* is exceptional in Iridaceae in having a superior ovary.

Selected Genera of Iridaceae
Babiana • Crocus • Cypella • Dierama • Dietes • Diplarrhena • Freesia • Gladiolus • Herbertia • Hesperantha • Iris • Ixia • Libertia • Moraea • Patersonia • Sisyrinchium • Sparaxis • Tigridia • Watsonia

JUGLANDACEAE

Walnut Family

About 60 species in 8 genera

RANGE Mainly northern hemisphere, cold and temperate, also South American Andes and tropical Malesia

PLANT FORM Mostly trees, a few shrubs, mostly deciduous; leaves in spirals (rarely opposite), pinnately compound or trifoliolate, leaflets nearly opposite, often toothed, without stipules, often aromatic; twigs are distinctive, having prominent leaf scars and sap clear but staining brown

FLOWERS Unisexual, small, with males in long catkins, usually wind pollinated; males numerous, in axils of catkin bracts, with perianth of (1 to) 4 or 5 tepals or perianth lacking; stamens mostly 5 to many, in 1 or 2 whorls, filaments short, anthers with longitudinal slits; females few, in clusters, with bracts often united forming an involucre, tepals toothlike or lacking; ovary inferior, of 2 or 3 united carpels with as many cells as carpels but with 1 ovule, styles separate and large, as many as carpels

FRUIT A large nut with a woody coat derived from the outer layer of the ovary and with a soft husk derived from the involucre, with a single seed, and embryo with convoluted cotyledons, or smaller with wings

An important family in North America, Juglandaceae comprise deciduous trees that are an important component of native forests. The flowers are unisexual and much reduced, the male flowers borne in long catkins with reduced perianth, and the females in smaller clusters. The leaves are compound in contrast to other families with male flowers in catkins, such as Betulaceae (alders, birches, hazels, hornbeams) and Fagaceae (beeches, chestnuts, oaks).

The genus *Carya* includes the hickories, the nuts of which are usually bitter and astringent (*C. cordiformis*, bitternut hickory; *C. glabra*, pignut hickory). Pecans, the nuts of *C. illinoinensis*, however, are sweet and edible, and much in demand in cooking (pecan pies), confectionary, or eaten as a snack. Many cultivars are available, mostly large trees grown in plantations mainly in southeastern North America but as far north as Illinois. Smaller cultivars are suitable for gardens. Juglandaceae also include the walnuts, *Juglans* species. The walnuts of commerce are the fruits of *J. regia* (English or European walnut), native to central Asia from Turkey to western China. The strongly flavored nuts of the North American *J. cinerea* (butternut) and *J. nigra* (black walnut) are mostly used in confectionary. Species of the genus *Pterocarya* (wingnut), with winged seeds unusual for the family, are grown as ornamentals, mainly *P. fraxinifolia* or its hardier hybrid with *P. stenoptera*.

Walnuts and hickories, especially black walnut (*Juglans nigra*), yield valuable hardwood, much used in furniture and cabinetry. Oil from seeds is used in paints, and a dye from husks is used in stains for flooring and wool. Other genera are also valued for their timber, and bark for tanning. Walnut trees produce compounds that inhibit the growth of some plants, especially species of Ericaceae, growing in their shade.

Genera of Juglandaceae

Alfaroa • Carya • Cyclocarya • Engelhardtia • Juglans • Oreomunnea • Platycarya • Pterocarya

Lamiaceae, *Salvia africana-lutea*. Typical of the family are the opposite leaves on squared or four-angled stems and the aromatic, gland-tipped hairs. The calyx persists around the fruit, and the flowers are strongly two-lipped. The ovary is deeply four-lobed, with the style inserted between the lobes and soon deciduous, and the stigma is two-lipped. *Salvia* is characterized by the two highly modified stamens in which the connective is hinged on a short, stout filament and markedly elongated so that the two anther lobes are well separated from one another. Only one of the anther lobes is fertile and is held under the upper floral lip on the arched connective.

LAMIACEAE

Mint or Sage Family

About 6700 species in 240 genera

RANGE Cosmopolitan

PLANT FORM Trees, shrubs, vines, and perennial herbs, a few annuals, stems of herbaceous species, especially when young, four-angled; leaves opposite or in whorls, usually simple or pinnately or digitately compound (especially *Vitex*), without stipules; containing essential oils and terpenes, thus aromatic

FLOWERS Perfect, usually zygomorphic and 2-lipped, in terminal cymes, spikes, or crowded in axils of leaves, sometimes solitary in axils; calyx mostly of 4 or 5 united sepals with short, rounded to tooth-like lobes, usually persisting around fruit; corolla mostly of 5 united petals, mostly 2-lipped, the lips unequal, rarely radially symmetric; stamens usually 4, or 2 plus 2 sterile stamens (staminodes), (or 5–8), inserted on corolla tube, anthers with longitudinal slits; ovary superior, of 2 united carpels, 2-celled, each carpel often divided in 2, the ovary thus 4-lobed and each lobe containing 1 ovule, with 1 style either terminal or basal, usually forked or 2-lobed at tip

FRUIT Usually four (or fewer) more or less separate one-seeded bony nutlets, when style basal the nutlets surrounding the more or less persistent style, or fruit fleshy

An important family well known in horticulture and as culinary herbs, Lamiaceae (historical name Labiatae) are recognizable by their familiar vegetative characteristics: squared stems bearing opposite or whorled leaves minutely dotted with oil glands and nearly always aromatic when crushed, and often with at least some flowers in the leaf axils. The flowers are also characteristic, especially the two-lipped corolla with two pairs of stamens combined with a four-lobed, superior ovary with a single style, this often basal or less often terminal. Labiates (Lamiaceae) are sometimes confused with borages (Boraginaceae), but the latter's coiled (scorpioid) inflorescence of mostly radially symmetric flowers with five stamens allows for easy separation. The verbenas (Verbenaceae) are superficially like labiates (and also aromatic) and are recognizable only with difficulty, typically by their terminal style, but this is also present in some Lamiaceae. Distinguishing between some Lamiaceae and Verbenaceae is simply not possible using morphology, and some authorities prefer to unite the two families.

With more than 800 species, *Salvia* (for example, *S. africana-lutea*, brown sage) is the largest genus of the family and is distributed almost worldwide. Among numerous ornamentals, the red salvia, *S. splendens*, native to Brazil, is probably the labiate most often seen in gardens and is now available in several different colors. Among the American species, *S. patens*, a Mexican native, has remarkable sky-blue flowers. *Salvia officinalis*, the herb sage, also has attractive blue

Lamiaceae, *Phlomis bracteosa*

Lamiaceae, *Phlomis lychnitis*

Lamiaceae, *Ajuga chamaepitys*

Lamiaceae, *Scutellaria prostrata*

to purple flowers and is a useful addition to any garden. Dozens more *Salvia* species are available in the nursery trade. Species of *Stachys* are also well-known garden plants, especially *Stachys byzantina*, lamb's ears, a reliable ground cover with softly hairy gray foliage. Another perennial species, *Stachys albotomentosa*, produces coral flowers throughout the summer and is a fine plant for the rock garden. The confusingly named North American sagebrush is *Artemisia tridentata*, a species of Asteraceae with aromatic, sage-scented foliage.

Species of *Phlomis* (especially *P. fruticosa*, Jerusalem sage) are attractive shrubs with velvety gray foliage and yellow flowers borne in clusters at nodes along the stems. The superficially similar African species *Leonotis leonurus*, with orange flowers, is hardy at least in Mediterranean climates and warm temperate gardens. Catmint (*Nepeta cataria*) and several more species of the genus *Nepeta* are equally or even more appealing plants for borders

or rock gardens. *Ajuga* is often used as a ground cover and has attractive blue flowers in spring. Any discussion of Lamiaceae dares not leave out lavender, the genus *Lavandula*, several species and cultivars of which are loved for their aromatic foliage and purple flowers borne in crowded terminal spikes. Species of *Lamium* are useful ground covers, especially the variegated cultivars of *L. maculatum*. *Moluccella*, bells of Ireland, with its enlarged green and persistent calyx, is often seen as a cut flower.

Among shrubs and small trees, *Vitex agnus-castus*, or simply vitex, is a striking plant with purple flowers. This and the peanut-butter tree, *Clerodendrum trichotomum*, were until more recently included in Verbenaceae. *Clerodendrum trichotomum* has attractive white flowers with a red calyx, contrasting with the fleshy blue fruit. Several more species of the genus are attractive shrubs, but unfortunately few are hardy. The Australian *Prostanthera* and *Westringia* are half-hardy shrubs useful in dry situations. Beauty

Lamiaceae, *Leonotus leonurus*

bush (*Callicarpa*) has insignificant flowers but remarkably blue or purple, berry-like fruit ripening in autumn. Many more Lamiaceae are familiar garden plants, including *Caryopteris*, *Monarda* (bee balm), *Perovskia* (Russian sage), *Plectranthus* (the variegated coleus of gardens is *P. scutellariodes*), *Scutellaria*, and *Teucrium* (germander).

Many more species of Lamiaceae are culinary herbs used in food preparation, and others are grown for their aromatic oils and for medicinal preparations. Culinary herbs include species of basil (*Ocimum basilicum*), mint (*Mentha*), oregano (*Origanum*), rosemary (*Rosmarinus*), thyme (*Thymus*), and savory (*Satureja*). The valuable hardwood teak is obtained from a tropical Asian species of *Tectona*.

Selected Genera of Lamiaceae

Agastache • Ajuga • Ballota • Basilicum • Calamintha • Callicarpa • Caryopteris • Clerodendrum • Clinopodium • Dracocephalum • Glechoma • Horminum • Hyssopus • Lallemantia • Lamium • Lavandula • Leonotis • Marrubium • Melissa • Melittis • Mentha • Moluccella • Monarda • Monardella • Nepeta • Ocimum • Origanum • Perilla • Perovskia • Phlomis • Physostegia • Plectranthus • Prunella • Pycnanthemum • Rosmarinus • Salvia • Satureja • Scutellaria • Sideritis • Solenostemon • Stachys • Teucrium • Thymus • Vitex • Westringia • Ziziphora

LAURACEAE

Laurel Family

About 2700 species in 56 genera

RANGE Nearly cosmopolitan but not cold temperate, mostly tropical South America and southeastern Asia

PLANT FORM Trees and shrubs, and a parasitic twining plant (*Cassytha*), the roots, stems, and leaves usually aromatic; leaves usually in spirals, rarely whorls, simple, entire or lobed (scale-like in *Cassytha*), without stipules

FLOWERS Usually bisexual, radially symmetric, axillary or in thyrses or umbels, usually small and 3-merous, with hypanthium tube; calyx and corolla not differentiated, 2 whorls of 3 green or yellow tepals each; stamens in 4 whorls of 3 each or inner whorl sterile or lacking, inserted at bases of tepals, anthers introrse, opening by terminal flaps (valves) closing when touched; ovary usually superior, usually 1 carpel, 1-celled with 1 large ovule, style 1, short or long

FRUIT A berry or drupe (for example, the avocado, *Persea*)

Lauraceae, *Laurus nobilis*

Few Lauraceae are hardy and hence of only modest importance to temperate horticulture. There are, however, some important exceptions. *Laurus nobilis*, the bay tree or bay laurel, is relatively hardy and certainly so in Mediterranean climates. A small to medium-sized evergreen tree with small yellow flowers, its aromatic leaves are used in cooking, either fresh or dried, as the bay leaf. North American *Sassafras* is likewise hardy and an attractive small tree, deciduous with good autumn color. The pulverized leaves of *Sassafras* provide the spice filé, used in the preparation of gumbo, the spicy stew of Cajun cuisine.

Some species of *Lindera* are also hardy and grown for their attractive foliage, turning bright red in autumn. The California laurel or olive, *Umbellularia*, also relatively hardy, is a handsome evergreen tree of western North America. Its aromatic leaves may be substituted for bay leaves but have a stronger flavor. The hardwood, sometimes misleadingly called myrtle

(true myrtle is *Myrtus* of the Myrtaceae), is used for furniture and carvings. Exceptional for the family, the genus *Cassytha* is an epiphytic parasite of trees and shrubs, with green stems and vestigial leaves.

Lauraceae include the avocado (*Persea americana*), which is hardy in Mediterranean climates and grown extensively in California, Chile, and Mediterranean Europe as well as in the tropics. Many Lauraceae provide valuable timber, and that of the southern African *Ocotea bullata* (stinkwood) is much prized for its hard, fragrant wood and also for its bark, which purportedly has medicinal properties. Many tropical Lauraceae also provide valuable timber. The genus *Cinnamomum* includes *C. verum*, source of the spice cinnamon; *C. aromaticum* is the source of false or bastard cinnamon, and camphor was originally obtained from *C. camphora*.

Lauraceae can be recognized by their simple but sometimes lobed, usually leathery leaves lacking stipules, tiny flowers with the perianth parts alike, with three tepals in each of two whorls, and tiny anthers opening by terminal flaps that close when touched (usually only visible under a hand lens). The fruit is fleshy and contains a single seed.

Selected Genera of Lauraceae
Bielschmiedia • *Cassytha* • *Cinnamomum* • *Cryptocarya* • *Laurus* • *Lindera* • *Litsea* • *Nectandra* • *Ocotea* • *Persea* • *Sassafras* • *Umbellularia*

LILIACEAE

Lily Family

About 550 species in 15 genera

RANGE Northern hemisphere and mostly temperate, some in tropical mountains

PLANT FORM Perennials with rhizomes or bulbs, some vines (*Smilax*), stems often unbranched, bearing leaves or not; leaves simple, in spirals or whorls, usually parallel veined

FLOWERS Perfect, radially symmetric or zygomorphic, in racemes, umbels, or solitary, sometimes in the axils of leaves; calyx and corolla not differentiated, thus both petal-like and usually similar in size and color, free or fused below, tepals 6 in 2 whorls, upright or nodding to pendent; stamens 6, sometimes arching downward over lower tepals, inserted at base of tepals, anthers extrorse; ovary superior, compound, of 3 united carpels, 3-celled, with a 1 terminal style, stigma sometimes 3-forked or single

FRUIT A dry capsule, often with the seeds flattened, seeds sometimes with white, fleshy appendage (an elaiosome)

The circumscription and composition of Liliaceae have changed radically since 1990, first with the removal of Asparagaceae and closely related families to the separate order Asparagales, which include families usually with black seed coats and introrse anthers among other features. Subsequently, additional genera were segregated into the families Colchicaceae (*Colchicum*, *Disporum*, *Uvularia*), Convallariaceae (*Convallaria*, *Polygonatum*), and Melanthiaceae (*Melanthium*, *Trillium*, *Veratrum*, and several others).

Liliaceae still include many much-loved garden subjects and cut flowers, importantly *Tulipa* (tulips), native to western and central Eurasia, *Fritillaria* (fritillaries) of the northern-hemisphere temperate zone, and of course the genus *Lilium* itself, with some 110 species (for example, *L. lanceolatum*, tiger lily).

Breeding and cultivation of lilies and tulips is a multimillion-dollar industry, both for the bulb and cut-flower trade. Today, hundreds of named cultivars of lilies are available in the nursery trade, and several true species are also grown, most cold hardy. Several species of *Erythronium* (dogtooth violets, glacier lilies), a genus allied to *Tulipa* but with nodding to pendent flowers, are also grown, more often in specialist collections. Seldom seen in cultivation, the western North American genus *Calochortus* includes a host of tulip-like species, many with striking flowers and deserving of cultivation.

No Liliaceae are poisonous, and bulbs of some *Lilium* species are eaten as a vegetable in China, a startling thought to those gardeners who treasure lilies. Mice and voles will also eat bulbs of lilies and tulips (and other Liliaceae) in the ground.

Liliaceae, *Lilium lanceolatum*. The narrow, whorled leaves are typical of many lilies. Lacking a distinction between calyx and corolla, the perianth parts are referred to as tepals rather than petals and in *Lilium* are conspicuous and often brightly colored. The three-parted flowers have six stamens and an ovary of three united carpels with a single style terminating in a globular stigma.

Liliaceae, *Fritillaria persica*

Liliaceae, *Tulipa aganensis*

Liliaceae, *Erythronium grandiflorum*

Liliaceae, *Fritillaria pudica*

Genera of Liliaceae

Amana • Calochortus • Cardiocrinum • Clintonia • Erythronium • Fritillaria • Gagea • Lilium • Llyodia • Medeola • Nomocharis • Scoliopus • Streptopus • Tricyrtis • Tulipa

ALSTROEMERIACEAE Plants in this family have most of the characteristics of Liliaceae but are easily distinguishable by their leaves, which are twisted at the base so that the upper surface is turned to face the ground. The rootstock is a rhizome, also found in some Liliaceae. *Alstromeria*, with nodding flowers, is widely grown. It and the closely related *Bomarea*, a genus of trailing to vine-like plants, have an inferior ovary and capsular fruits. In contrast, the southern South American and New Zealand *Luzuriaga* and related Australasian *Drymophila*, both somewhat woody vines, have a superior ovary, and the fruit is a berry.

LYTHRACEAE

Loosestrife or Crepe Myrtle Family

About 500 species in 28 genera

RANGE Nearly cosmopolitan but mostly tropical and sub-tropical, a few temperate

PLANT FORM A few trees (including mangroves, trees growing in tidal saline water along coasts, for example, *Sonneratia*) with hard wood and often flaking bark, some shrubs, many perennial herbs, and a few annuals, including aquatics; leaves opposite, sometimes in whorls or spirals, simple, stipules vestigial or lacking

FLOWERS Perfect, often heterostylous [stamens and style at different levels in different plants, thus stamens long and style(s) short versus stamens short and style(s) long], in racemes, axillary clusters, or solitary, with floral cup (hypanthium), radially symmetric or zygomorphic; calyx of 4 (or 5–8) free sepals; corolla of free petals as many as sepals, inserted at or below mouth of hypanthium, crumpled in bud; stamens usually twice as many as petals, in 2 whorls (rarely fewer or many), inserted on hypanthium, anthers with longitudinal slits; ovary mostly superior but inferior in *Punica*, of 2–4(–6) united carpels, many-celled, with 1 style

FRUIT Usually a dry capsule, sometimes more or less leathery, usually with small, dry seeds but those of *Punica* (pomegranate) with a fleshy, edible seed coat

Lythraceae contribute little to temperate horticulture, although with global warming the area that supports members of the family is increasing. *Lagerstroemia indica* (crepe myrtle or Pride of India), one of several trees in the family, is grown across the southern United States, in California, the Mediterranean, Australia, and southern Africa. Now it can also be grown in parts of the North American Midwest where it did not survive before 1990. Petals inserted at or below the mouth of the floral cup and obviously crumpled in bud make Lythraceae easy to recognize. Other members of the family in cultivation are *Lythrum* (loosestrife) and several species of *Cuphea*, treated as annuals in temperate gardens.

Several cultivars of *Punica*, usually with brilliant orange flowers, are hardy at least in warm temperate and Mediterranean climates and are used as hedges as well as ornamentals and grown in orchards. Older classifications treat *Punica* as a separate family, Punicaceae, but molecular studies render the recognition of Punicaceae redundant.

Lythraceae contribute modestly to the list of useful or edible plants, but the pomegranate (*Punica granatum*) is a notable exception, providing fresh fruit in dry parts of the Old World and now widely cultivated. The pomegranate has been used by humans for more than 10,000 years, and its importance is reflected in the pomegranate motif adorning buildings across ancient Egypt and Mesopotamia. The eastern Mediterranean *Lawsonia inermis* is the source of the red dye henna, used as a hair coloring, for temporary tattoos, and for cloth. Henna is known to have been used in ancient Egypt some 6000 years ago. *Lythrum salicaria* (purple loosestrife) native in Eurasia, has become an invasive weed in wetlands in North America, degrading valuable habitat for wildlife.

Selected Genera of Lythraceae
Ammania • Cuphea • Lafoensia • Lagerstroemia • Lawsonia • Lythrum • Nesaea • Pemphis • Punica • Rotala • Sonneratia • Trapa • Woodfordia

MAGNOLIACEAE

Magnolia Family

About 220 species in 2 genera

RANGE Mostly northern hemisphere, also South America, temperate and tropical

PLANT FORM Trees and shrubs; leaves in spirals, entire or lobed, with large stipules enclosing terminal buds, falling as leaves unfold

FLOWERS Perfect, usually radially symmetric, large, solitary, and terminal on branches, usually with long receptacle; perianth not differentiated into calyx and corolla, thus consisting of 6–18 free tepals, usually arranged in spirals; stamens many in spiral arrangement, anthers with longitudinal slits; carpels few to many, free, crowded in spirals, each with 1 style and stigma, 1-celled, with 1 to few ovules

FRUIT Dry follicles (sometimes not splitting at maturity) or carpels growing together into a fleshy compound fruit, or dry and winged; seeds often red and when released suspended by fine threads

Although treated in the past as separate genera, molecular studies show that *Kmeria*, *Manglietia*, *Michelia*, *Pachylarnax*, and *Talauma* are nested within *Magnolia*, and their species have accordingly been included in that genus. For convenience, these former genera may informally be referred to as groups within *Magnolia*, which now includes all but two of the species in the family, the other genus being *Liriodendron*. Many *Magnolia* species, both evergreen and deciduous, are hardy, and several are important for ornamental horticulture. Most widely cultivated of the evergreen species, *M. grandiflora* (loblolly, southern magnolia, bull bay) is native to southeastern North America and produces dinner-plate-sized fragrant flowers, and glossy leaves usually with rusty brown tomentum on the underside. Although typically a large tree, several small or even dwarf cultivars have been selected that are suitable for the small garden.

Deciduous species include *Magnolia macrophylla* (big-leaf magnolia), which has the largest leaves of any native North American plant and flowers even larger than those of *M. grandiflora*. One of the most common magnolias in cultivation, *M. ×soulangeana*, is an old hybrid, created circa 1820. *Magnolia stellata* (star magnolia), one of the smaller species of the genus, has white or pink blooms with numerous, narrow tepals. *Magnolia acuminata*, known as the cucumber tree for its elongate and slightly fleshy compound fruit, has the largest flowers of the deciduous species, rivaling those of *M. grandiflora*. Unusual in the genus, *M. sieboldii* and *M. watsonii* have pendent white flowers with red stamens, also pleasantly scented. Although flowers of most *Magnolia* species are pleasantly fragrant, those of *M.* (*Michelia*) *figo* have an extraordinary and complex odor of bubble gum, coconut, and pineapple. Magnoliaceae are one of the oldest and most primitive families of flowering plants; fossils attributed to the family have been found in Cretaceous deposits more than 320 million years old.

Magnoliaceae, *Magnolia stellata*. Lacking a true calyx, magnolia flowers consist of many free, petal-like tepals in a spiral arrangement, numerous stamens spirally arranged, and several free but closely crowded carpels. The flowers are enclosed in bud by two or more somewhat furry bracts that are shed as soon as the flower opens.

Magnoliaceae, *Magnolia laevifolia*

Magnoliaceae, *Magnolia stellata*

Magnoliaceae, *Magnolia sieboldii*

The second genus of Magnoliaceae, *Liriodendron* (tulip tree) has just two species: *L. tulipifera* of eastern North America and *L. chinense* of China and Indochina. Both in cultivation, they are hardy and provide hardwood for furniture, siding, shingles, and building material. They are long lived and can grow to immense size. Some *Magnolia* species also provide useful timber and essential oils for perfume, and some have medicinal properties.

Magnoliaceae are recognizable by their simple, stipulate leaves arranged in spirals, terminal flowers lacking distinction between calyx and corolla, and the often large tepals arranged in spirals on an elongate receptacle. The numerous stamens and free carpels are likewise arranged in spirals. The stipules cover and protect the young leaves and are shed as the leaves unfold.

Genera of Magnoliaceae
Liriodendron • *Magnolia,* including *Kmeria, Manglietia, Michelia, Pachylarnax,* and *Talauma*

CALYCANTHACEAE With just three genera of North America and eastern Asia, Calycanthaceae have most of the diagnostic features of Magnoliaceae, to which they are closely related. Best known is *Calycanthus*, which includes *C. floridus* (Carolina allspice), *C. occidentalis* (California allspice), and *C. chinensis* (Chinese sweetshrub), the last sometimes placed in the genus *Sinocalycanthus.* (The spice pimento or allspice, used to flavor meat dishes and deserts, is obtained from *Pimenta officinalis* in the family Myrtaceae.) The bark has an aromatic, camphor-like odor. Winter flowering but hardy and occasionally found in gardens, *Chimonanthus* (wintersweet) has nodding white flowers with a delicious sweet scent. Like Magnoliaceae, the terminal, solitary or axillary flowers lack a calyx and have numerous tepals and 5–20 stamens arranged in spirals. The leaves are opposite and lack stipules. The carpels are inserted on a floral cup and contain a single ovule but do not split open, the fruit thus an achene. The fruit in most Magnoliaceae is a follicle, thus splitting to release the seeds.

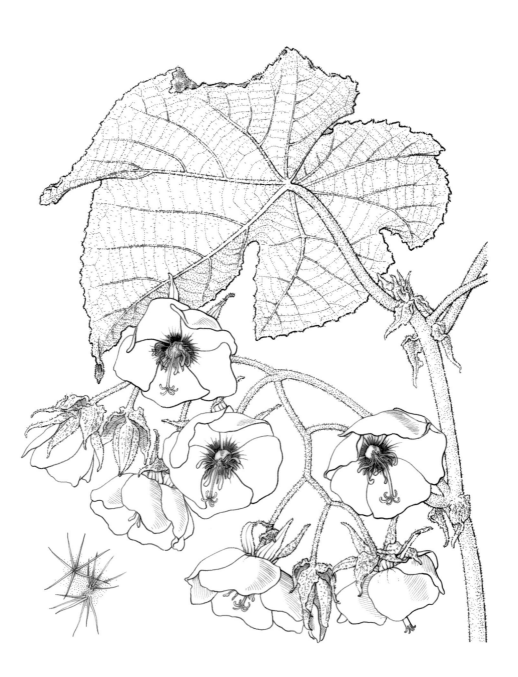

Malvaceae, *Dombeya burgessiae*, pink dombeya or pink wild pear. Although the stipulate, palmate leaves of *D. burgessiae* are typical of many Malvaceae, the diagnostic feature of the family is the harsh, star-like hairs that cover leaves and stems. The attractive, radially symmetric flowers have a five-lobed calyx subtended by an epicalyx of bracts, and five free petals that are characteristically furled in bud. In *Dombeya*, the numerous stamens are united in clusters at the base into a short tube alternating with five sterile staminodes.

MALVACEAE

Mallow or Hibiscus Family

About 5000 species in 230 genera

RANGE Cosmopolitan, especially tropical

PLANT FORM Trees, shrubs, perennial herbs, and some annuals, often with star-like hairs; leaves in spirals, entire to lobed or dissected, often palmately veined, or digitately compound, usually with stipules

FLOWERS Perfect, usually radially symmetric, in cymes or solitary in axils, usually with an epicalyx (a ring of 3 or more bracts below the flower); calyx of 5 sepals, sometimes basally united; corolla usually of 5 free petals (sometimes attached to base of staminal column); stamens 5 to many, often joined below in a tubular column or in clusters of 5 but the filaments free distally, anthers with longitudinal slits; ovary superior, of (1 to) 5 to many united carpels, 5- to many-celled with 1 to many ovules, styles as many as carpels

FRUIT Capsules, berries, or schizocarps, splitting into sections when ripe

As a result of molecular study of Malvaceae and related families, it is impossible to uphold recognition of the separate families Bombacaceae (baobab family), Sterculiaceae, and Tiliaceae (linden family), and their genera are now included within an expanded Malvaceae. Most of the many genera of these families are tropical and of no concern to temperate horticulture. In its broader circumscription, Malvaceae now include some surprising additions such as *Tilia* (the linden or lime tree). Few *Hibiscus* species are hardy, and some of the most attractive may need to be overwintered in a greenhouse. Nevertheless, species such as *H. syriacus* (rose of Sharon) are usually hardy. Many other Malvaceae grown for display in gardens are perennials, including several species of *Abutilon* (Indian mallow), *Alcea* (hollyhock), *Lavatera* and *Malva* (mallows), and the North American *Sidalcea* (checker flower, prairie mallow). The Chinese umbrella tree (*Firmiana*), hardy in areas of Mediterranean climate and the coastal temperate zone, is particularly striking with large leaves crowded at the branch tips, recalling an umbrella. The yellow-flowered California native *Fremontodendron* is one of the most attractive shrubs in the family.

Malvaceae are of considerable economic importance for timber, fiber, and twine, most notably species of *Gossypium*, which provide cotton derived from fine hairs covering the seeds. Jute, used for rope, cloth bags, and carpeting, is a fiber obtained from a species of *Corchorus*. Several genera provide food products, including *Theobroma*, source of cocoa and thus chocolate, and *Cola*, the nuts of which provide the flavoring of that name. The durian, fruit of the tree *Durio zibethinus*, is eaten avidly by humans and other animals in Southeast Asia and adjacent areas. Among vegetables, okra, the young fruits of which are eaten fried or stewed (in gumbo), is *Hibiscus esculentus* (in the past placed in the genus *Abelmoschus*).

The family is recognizable by its five-merous flowers, free or basally united sepals, often the presence of a filament column with multiple stamens to which free petals are often attached, and a superior ovary with as many styles as carpels. The leaves, at least of those of temperate-zone species, are simple, often with palmate venation. *Sterculia* and several other genera have digitately compound leaves.

Selected Genera of Malvaceae

Abutilon • Adansonia • Alcea • Althaea • Callirhoe • Corchorus • Firmiana • Fremontodendron • Gossypium • Hermannia • Hibiscus • Lavatera • Malva • Malvastrum • Pavonia • Sida • Sidalcea • Sparrmannia • Sphaeralcea • Sterculia • Tilia • Theobroma

Malvaceae, *Hibiscus trionum*

Malvaceae, *Anisodontea juliae*

MELANTHIACEAE

Melanthium Family

About 190 species in 17 genera

RANGE Northern hemisphere to tropical Asia and South America

PLANT FORM Perennials with condensed rhizomes or sometimes bulbs, stems branched or exclusively unbranched, bearing leaves or not; leaves in spirals or in two ranks, usually parallel veined, but in a whorl of four leaves with reticulate venation in *Paris* and three in *Trillium*

FLOWERS Perfect, radially symmetric, in racemes, spikes, solitary in *Paris* and *Trillium*, upright or nodding to pendent; calyx and corolla usually not differentiated, thus both whorls petal-like and usually similar in size and color (sometimes outer whorl small, green, and calyx-like outside, notably in *Trillium*), free or joined below, tepals usually 6 in 2 whorls (4-merous in *Paris*), often persisting in fruit; stamens 6, inserted at base of tepals, anthers extrorse; ovary superior, compound, of 3 united carpels, 3-celled, often with separate styles and stigmas

FRUIT A dry capsule or fleshy and berry-like; seeds often with fleshy, white appendage (an elaiosome) and dispersed by ants

Melanthiaceae include only a few plants commonly seen in gardens, but *Trillium* species are sometimes cultivated and are conspicuous in woodlands in North America in spring; the common name wake robin reflects its early blooming. A fairly large temperate Eurasian and North American genus, *Trillium* is characterized by its solitary whorl of three leaves, the stalks of which are fused together, and includes species with stalked or sessile flowers. *Paris*, a close relative of *Trillium* with rather curious looking but not exactly beautiful flowers, is seen less often in gardens. Gardens emphasizing native plants grow one or more species of *Melanthium*, *Veratrum*, or *Zigadenus*.

Many species of these genera are used medicinally and may be highly toxic, hence one of the common

Melanthiaceae, *Trillium sessiliflorum*

Melanthiaceae, *Trillium erectum*

names is death camas. Some scientific names, likewise, denote the poisonous nature of species such as *Toxicoscordion venenosum*, "venomous poison plant." Toxic principles include steroidal alkaloids that can depress heart rate, resulting in low blood pressure, or cause cardiac arrhythmia. Happily, *Trillium* species are not dangerous when ingested, but the related genus *Paris* is reported to be poisonous, especially the fruits. Usually avoided by stock, consumption of some Melanthiaceae can result in birth deformities in sheep.

Melanthiaceae are closely related to Liliaceae and in the past were included in that family but are distinguishable by their rootstock, often a rhizome (or a bulb), petals and sepals often persisting in fruit, and carpels usually with separate styles. Many genera contain highly toxic compounds unknown in true Liliaceae.

Selected Genera of Melanthiaceae
Amianthum • Anticlea • Chamaelirion • Helonias • Paris • Stenanthium • Toxicoscordion • Trillium • Veratrum • Xerophyllum • Zigadenus

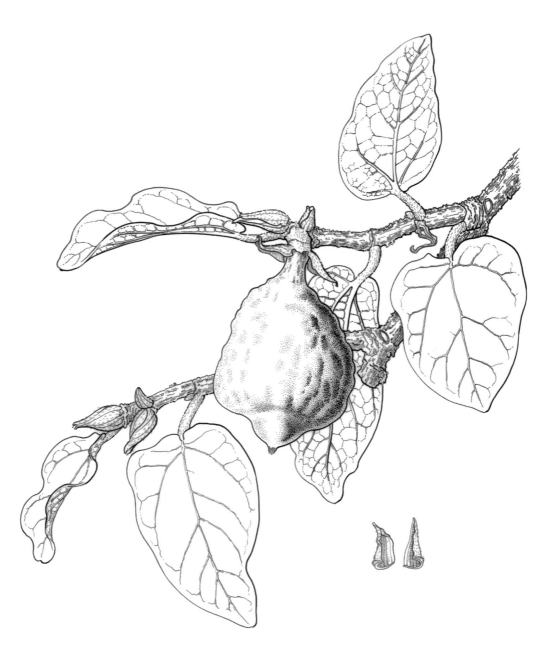

Moraceae, *Ficus pumila*. This evergreen creeper covers vertical surfaces with a tracery of fine stems densely covered with small, heart-shaped leaves, creating a mat of foliage. These are the juvenile leaves. Once the vine reaches the top of its support it will begin to form horizontal branches with larger adult leaves. Milky sap and large deciduous stipules enclosing the apical bud are both typical of the family. The distinctively shaped figs enclosing numerous minute flowers are borne only on the horizontal stems.

MORACEAE

Mulberry Family

About 1200 species in 40 genera

RANGE Mostly tropics and subtropics, few temperate

PLANT FORM Trees, shrubs, vines, and some perennial herbs, usually with milky latex; leaves usually simple, sometimes lobed, in spirals or opposite, leathery, usually with stipules, these sometimes minute

FLOWERS Perfect or unisexual, small, in axillary inflorescences, the axis often thickened to form a head, in *Ficus* enlarged to enclose flowers, with a small opening at the tip; calyx and corolla not differentiated, thus perianth with 0–10 tepals, sometimes united basally; stamens 1–6, as many as tepals, partially united; ovary superior or inferior, of 2 or 3 united carpels, 1-celled and with 1 ovule, usually with 2 styles or 1 style deeply divided

FRUIT Compound, consisting of multiple drupes (each one-seeded), thus *Morus* (mulberries, which are not true berries) or *Ficus* (figs)

Moraceae offer little to temperate horticulture, but two genera provide edible fruit. Species of *Morus* (mulberry) are hardy, especially the white mulberry, which is widely grown in the Middle East. *Ficus* (figs, for example, *F. pumila*, creeping fig), with some 850 species, is the largest genus in the family but almost exclusively tropical. *Ficus carica* (common fig) is hardy in Mediterranean climates and sometimes in coastal temperate zones. There are numerous cultivars of figs, some producing fruit without pollination, others requiring pollination by a complex strategy using minute fig wasps that reproduce within the developing inflorescence. Fruits of tropical species are important as food for many mammals and birds, but few are significant for humans. Several fig species are grown as indoor plants or maintained in greenhouses in winter, especially *F. benjamina* (*F. nitida*), which is hardy in Mediterranean climates. The Osage orange, *Maclura pomifera* of North America, is one of few cold-hardy Moraceae and is occasionally cultivated in the Midwestern United States.

The family is of moderate economic importance, but *Artocarpus* provides breadfruit and jackfruit, important food sources in some tropical cultures. Leaves of *Morus alba*, white mulberry, are the main food of silkworms (the larvae of the silk moth, *Bombyx mori*, the cocoons of which are the source of silk) though they can eat some other species. Several tropical Moraceae provide timber.

Selected Genera of Moraceae
Artocarpus • *Dorstenia* • *Ficus* • *Maclura* • *Morus* • *Naucleopsis*

Myrtaceae, *Eugenia myrtifolia. Eugenia* has the opposite leaves typical of the family as well as flowers with sepals and relatively inconspicuous petals inserted on a hypanthium. As in most Myrtaceae, the numerous, prominent, colorful stamens serve to attract pollinators. The fruit is fleshy and retains the remains of the hypanthium and calyx.

MYRTACEAE

Myrtle or Eucalyptus Family

About 5500 species in 130 genera

RANGE Mostly tropics and warm temperate, especially Australia

PLANT FORM Trees, shrubs, vines, and some herbs, usually with abundant aromatic oils; leaves simple, mostly opposite, leathery, usually gland dotted, usually lacking stipules

FLOWERS Usually perfect, with hypanthium extending above ovary, in compound inflorescences, rarely solitary (*Myrtus*); calyx mostly of 4 or 5 free sepals, sometimes lobes separating at flowering or forming a cap (calyptra), falling as the flower opens; corolla of 4 or 5 petals or lacking; stamens usually many, inserted on rim of hypanthium, often in clusters opposite the petals, anthers with longitudinal slits, rarely with pores, free or united in groups; ovary usually inferior, of 2–5(–16) carpels with as many cells, usually with many ovules in each, style 1

FRUIT A capsule or fleshy, either many-seeded, thus a berry, or a one-seeded drupe, rarely a nut

Although few Myrtaceae are cold hardy, many thrive in warm temperate and Mediterranean climates. The best-known genus is *Eucalyptus* (often called blue gum or simply gum) with about 680 species, including small to large trees, often with drooping foliage. Most have juvenile leaves that are surprisingly different in shape, color, and position from the adult leaves, with a waxy surface. Several species are grown in plantations for timber around the world and are sometimes invasive, especially in southern Africa. Among the most hardy species are *E. gunnii* (cider gum) and *E. niphophila;* they and several other species are grown as street trees in Australia and California. *Corymbia ficifolia* (previously *E. ficifolius*) has large inflorescences of pink to orange or red flowers, making it particularly ornamental.

The Mediterranean genus *Myrtus* (myrtle) is an attractive shrub with small, scented flowers. Historically, myrtle was used in religious rituals and as garlands in classical Rome. *Eugenia* (for example, *E. myrtifolia*, cherry eugenia) another large genus of trees and shrubs of Myrtaceae, includes some moderately hardy species, including the Surinam cherry or pitanga, *E. uniflora*. Its fleshy fruit is used for jellies and conserves. Species of the Australian genus *Callistemon* (bottlebrush) are remarkably attractive shrubs for gardens, the sessile flowers arranged in columnar, brush-like inflorescences. The apex grows beyond the flowers to produce new leaves and stems, an unusual growth pattern. Some species of *Leptospermum*, also Australian, are attractive shrubs, also used as hedges but unfortunately sometimes invasive. *Verticordia*, another Australian shrub, has especially striking flowers.

Myrtaceae, *Corymbia ficifolia*

Many Myrtaceae are of economic importance, particularly for their timber in Australia but elsewhere as well. The African and Pacific genus *Syzygium* includes *S. aromaticum* (formerly *Eugenia caryophyllata*), the source of cloves, the principal export of Zanzibar. Cloves, the dried flower buds of *S. aromaticum*, are an important flavoring for foods and have antibacterial, antiviral, and analgesic properties, commonly used for relief of toothache. Allspice or pimento is another culinary product, obtained from another tropical species, *Pimenta officinalis*. Mildly frost tolerant, species of *Acca* (better known under the name *Feijoa*) provide edible, guava-like fruits, and some have fleshy, edible flowers. The tropical American tree *Psidium guava* provides delicious fruit, the guava, technically a berry, now grown in orchards in the tropics and areas of Mediterranean climate.

Myrtaceae are most easily recognizable by their opposite leaves that are aromatic and glandular punctate (with translucent or colored dots or depressions). The flowers are often attractive, with a prominent brush of stamens borne on the rim of a hypanthium.

Selected Genera of Myrtaceae
Acca, including *Feijoa* • *Agonis* • *Angophora* • *Callistemon* • *Chamaelaucium* • *Darwinia* • *Eucalyptus* • *Eugenia* • *Leptospermum* • *Melaleuca* • *Metrosideros* • *Myrtus* • *Pimenta* • *Psidium* • *Syzygium* • *Verticordia*

Nyctaginaceae, *Mirabilis jalapa*. Nyctaginaceae are one of the families containing betalain pigments, which produce particularly brilliant purple or cerise colors. The simple leaves lack stipules, and the radially symmetric flowers are subtended by leafy bracts that resemble a calyx. The actual perianth comprises a single whorl of mostly five sepals joined below into a slender tube, with the lobes often pleated longitudinally in bud. The ovary contains a single ovule and ripens into a hard achene or nut.

NYCTAGINACEAE

Four-o'Clock or Bougainvillea Family

About 390 species in 27 genera

RANGE Tropics and subtropics, a few temperate, especially Americas

PLANT FORM Trees, shrubs, or perennial herbs, a few annuals; leaves simple, entire, in spirals or opposite, then sometimes unequal, without stipules

FLOWERS Bisexual, radially symmetric, often in cymes, often subtended by large colored bracts; perianth of 1 whorl, tubular, with (3–)5(–8) lobes often pleated longitudinally; stamens usually as many as perianth lobes or 1–40, filaments often unequal, anthers with longitudinal slits; ovary superior, of 1 carpel with 1 ovule, style long and slender

FRUIT A dry achene or nut

Of little importance in temperate horticulture, Nyctaginaceae include just a few ornamentals, most notably the scrambling vine *Bougainvillea*, native in South America. Usually spiny and with leaves in spirals, it has small, inconspicuous flowers that are dwarfed by large colored bracts that resemble parts of a flower. Mostly grown in the tropics, bougainvilleas are also often seen in areas with a Mediterranean climate and will withstand light frost, then often becoming deciduous. *Mirabilis jalapa*, the four-o'clock flower or marvel of Peru, is a tuberous-rooted perennial with large red, orange, or yellow flowers, opening in late afternoon and lasting a single day, collapsing the following morning. The sand verbenas, *Abronia* species, are low-growing and sprawling perennials or annuals of northern Mexico and the United States, with large clusters of flowers. Several species make attractive garden plants for dry areas, especially *A. latifolia* and *A. villosa*, the latter carpeting sandy flats with purple flowers in interior California after good winter rain.

Selected Genera of Nyctaginaceae
Abronia • Boerhavia • Bougainvillea • Mirabilis • Neea • Nyctaginia • Pisonia

PHYTOLACCACEAE Related to Nyctaginaceae, some plants in the pokeweed family are occasionally grown in temperate gardens but include no significant ornamentals. *Phytolacca* species include the potherb *P. americana* (pokeweed or pigeonberry), which also yields a dye to color ink, and poisonous plants including *P. dodecandra*. Leaves of the family are spirally arranged, the flowers have an ovary of (one to) several carpels with as many styles, and the fruits may be fleshy or dry and often separate into sections when ripe.

Nymphaeaceae, *Nymphaea nouchali*. Native to Asia and Australia, this ornamental water lily, like others in the family, has numerous petals and stamens arranged in spirals, and many free carpels. The leaves float on the surface of the water, but the flowers are held on long stalks above the surface.

NYMPHAEACEAE

Water Lily Family

About 75 species in 5 genera

RANGE Cosmopolitan

PLANT FORM Aquatic perennial herbs with rhizomes or tubers and floating leaves; leaves in spirals, round to heart-shaped, floating, with or without stipules

FLOWERS Bisexual, floating or aerial, solitary in axils; calyx mostly of 4–6 free sepals, sometimes petaloid or green; corolla (0 to) 8 to many petals, free, often grading into stamens; stamens many in spirals, usually free, petal-like or with filaments, anthers with longitudinal slits; ovary of 3 to many carpels, inferior or superior, with many ovules

FRUIT Fleshy, berry-like; seeds usually with fleshy appendages (not in *Nuphar*)

Consisting only of rooted aquatic herbs, Nymphaeaceae are mostly tropical, but the genus *Nuphar* (sometimes called pond lily) is hardy and native to temperate, even cold temperate climates. Often seen in ponds in gardens and lakes in parks, the true water lilies belong to the cosmopolitan genus *Nymphaea*, species and cultivars of which are widely grown (for example, *N. nouchali*, blue water lily or blue lotus, which served as the state flower of Hyderabad, India, and is the national flower of Bangladesh and Sri Lanka). Not cold hardy, plants are raised in containers in the temperate zone and moved indoors to overwinter. The South American genus *Victoria* has spines on the stems, petioles, and undersides of the leaves, and *V. amazonica*, giant water lily, has huge pads up to more than 6 feet (2 meters) in diameter that can support the weight of a child. Occasionally grown outdoors in gardens, plants are also brought indoors in winter. The dinner-plate-sized flowers open in the evening to display white petals, which gradually turn pink the following day.

Genera of Nymphaeaceae

Barklya · Euryale · Nuphar · Nymphaea · Victoria

NELUMBONACEAE Often associated with Nymphaeaceae, the sole genus *Nelumbo* is aquatic, including two species: *Nelumbo nucifera* (the sacred lotus) and *N. lutea* (American lotus). The Nelumbonaceae are now understood by molecular studies to be only distantly related to the true water lilies and instead are most likely closely allied to Proteaceae! These rhizome-bearing plants have large, water-lily-like, but emergent flowers with multiple tepals and stamens arranged in spirals. The umbrella-like leaves emerge above the water surface, as do the flowers and fruits, which are borne on long stalks and contain large, edible, nut-like seeds encased in distinctive, flat-topped, enlarged receptacles. *Nelumbo nucifera* is revered in eastern Asia, and numerous cultivars are maintained in gardens and parks. An emblematic plant, it is illustrated in many artworks and is a decorative theme in architecture.

Oleaceae, *Jasminum officinale*. The twining *J. officinale* is typical
of Oleaceae in the four- or five-parted calyx and corolla, the petals
of which are united below, often in a narrow tube. Flowers have a
superior ovary and almost invariably only two stamens usually hidden
in the floral tube. Like many members of the family, the flowers are
sweetly scented.

OLEACEAE

Olive or Jasmine Family

About 565 species in 25 genera

RANGE More or less worldwide with a center in eastern Asia but many in northern temperate zone, some tropical

PLANT FORM Mostly trees, shrubs, and subshrubs, some vines, evergreen or deciduous, stems usually four-angled, especially when young; leaves mostly opposite (rarely in spirals or whorls), simple or compound, pinnate or trifoliolate, without stipules

FLOWERS Perfect, radially symmetric, usually in stalked clusters in the axils of leaves or terminal, strongly scented in many species; calyx with sepals partly united, often tubular, 4- or 5-lobed or 4-toothed, rarely obsolete, usually green; corolla with petals partly united, 4- or 5-lobed, rarely obsolete (*Fraxinus*, true ash); stamens 2 (rarely 4 in genera not in cultivation), inserted on the corolla tube; ovary superior, compound, of 2 closely united carpels, thus 2-celled, with 1 style terminal, stigma 1- or 2-lobed, each carpel often with 1 ovule

FRUIT Often one-seeded drupes (for example, olives, *Olea*), or dry with a single wing (*Fraxinus*, true ash), also berries or capsules (*Syringa*, lilac)

Oleaceae are an important family of woody ornamental trees, shrubs, and vines, most notably true lilacs (*Syringa*), fringe trees (*Chionanthus*), *Forsythia*, and *Osmanthus* (sometimes called sweet olive; the generic name translates as fragrant flower). Some jasmines (*Jasminum*) are scrambling shrubs or vines. The so-called white forsythia is the genus *Abeliophyllum*, one of few genera endemic to Korea. Flowers of Oleaceae are often pleasantly scented, most notably the lilacs, *Abeliophyllum*, and many *Jasminum* and *Osmanthus* species. *Jasminum officinale* (common jasmine or poet's jasmine) has been grown for centuries in China, where it is native, and is half-hardy in North American gardens but thrives outdoors in Australia and South Africa as well as in other countries with a Mediterranean climate. Winter-flowering *J. nudiflorum* is a valued garden plant, one of few producing flowers in winter months. The relatively small lilac genus *Syringa* includes the hardiest members of the family, grown both for their floral display of pink, mauve, or white flowers and delightful fragrance. *Syringa* is closely related to the genus *Ligustrum* (privet), some evergreen species of which are used as hedges and topiary.

Some Oleaceae are economically important, especially the cultivated olive (*Olea europaea*), used for oil production; the fleshy fruit is preserved in various ways after leaching out the bitter components. Several species of true ash (*Fraxinus*) and *Ligustrum* are used for timber and urban forestry as street plantings. Some *Fraxinus* species are exceptional in Oleaceae in having inconspicuous flowers borne in pendent catkin-like racemes, and winged fruits. *Jasminum officinale* flowers are used in scent production, as are other species of the genus. Dried flowers of *J. sambac* are used to scent and flavor jasmine tea.

The family is recognizable by its opposite leaves, either simple or compound, relatively small flowers with the corolla typically of four lobes, united basally or in a narrow tube, mostly just two stamens, and a superior ovary with single terminal style. The single-seeded fleshy fruit, usually black, is also distinctive, but lilacs (*Syringa*) have dry capsules, and *Fraxinus*, with dry, winged fruits, is exceptional in Oleaceae.

Selected Genera of Oleaceae
Abeliophyllum • Chionanthus • Forsythia • Fraxinus • Jasminum • Ligustrum • Nestegis • Olea • Osmanthus • Parasyringa • Schrebera • Syringa

Oleaceae, *Syringa vulgaris*

Oleaceae, *Jasminum multipartitum*

ONAGRACEAE

Evening Primrose Family

About 650 species in 21 genera

RANGE Mainly North, Central, and South America, also eastern Asia, New Zealand, poorly represented elsewhere in tropics

PLANT FORM Shrubs, a few trees, many perennials and annuals; leaves opposite or in spirals, sometimes in whorls or all basal, simple, entire, toothed, or lobed, stipules present and soon falling or absent

FLOWERS Usually bisexual (rarely unisexual), usually radially symmetric (not *Epilobium* section *Zauschneria*), in spikes, sometimes solitary, with a floral tube (hypanthium); sepals usually 4 (or 2, 3, 5–7), sometimes brightly colored, inserted at mouth of hypanthium; petals as many as sepals; stamens usually twice as many as petals in 2 whorls, sometimes only as many as petals, rarely 2 (*Circaea, Lopezia*); ovary inferior, of 4 united carpels, usually 4-celled, often with many ovules, style terminal, 4-lobed at apex or undivided

FRUIT Usually a capsule containing many seeds, or a berry (*Fuchsia*) or hard and nut-like

Onagraceae include numerous species of garden plants, perhaps most notably those of *Gaura* (now a section of the genus *Oenothera*), widely grown in temperate and subtropical areas of both hemispheres for their long flowering season and cheerful white or pink, butterfly-like blooms. Several species and cultivars of *Oenothera* are also often grown for their large yellow or pink flowers, although several others are weedy in many parts of the world. More widely cultivated in the tropics and warm temperate zones is the genus *Fuchsia*, of which both wild species and numerous large-flowered hybrids are available in the nursery trade. Few species are hardy in cold temperate climates, but the small-flowered *F. magellanica* is an exception; its cultivars and hybrids with the species survive fairly cold winter conditions. Many more cultivars of *Fuchsia* are either grown as annuals or under glass.

Of the genus *Epilobium* with more than 160 species, often known as willow herb, only the so-called California fuchsias are in general cultivation. Known in the past as the genus *Zauschneria*, these Californian species with long-tubed, red flowers are adapted for pollination by hummingbirds but otherwise have the attributes of *Epilobium*, including the distinctive elongated seeds with a terminal tuft of fine, silky hairs. Flowering from late summer to late autumn, *E. canum* and *E. septentrionalis* make stunning cushions of scarlet blooms among the gray or sometimes pale green foliage. Most other species of *Epilobium* have inconspicuous white to pink flowers, and several are notorious garden weeds. Cushion-forming species from New Zealand are sometimes grown as alpines in rock gardens. With striking, large, purple flowers, the northern temperate fireweed or rosebay willowherb has been removed from *Epilobium* to the genus *Chamaenerion* as *C. angustifolium*.

Onagraceae, *Fuchsia fulgens.* Typical of Onagraceae, the flowers of fuchsia are four-merous, with a calyx of prominent, colorful sepals surrounding the petals and joined to them to form a tubular hypanthium. The stamens are twice as many as the petals, and the ovary has a solitary style with a four-lobed stigma.

Onagraceae, *Chamaenerion latifolium*

Onagraceae, *Epilobium canum*

Onagraceae, *Ludwigia octovalvis*

During the Second World War, the species took root and flourished on many of the London bomb sites and commonly came to be called fireweed in memory of the blitz. Some annuals of the genus *Clarkia*, native to western North America and mostly California, are fine garden subjects but today only occasionally cultivated. *Ludwigia* species are popular aquarium plants and sometimes grown around ponds.

Most Onagraceae can readily be recognized by the presence of a floral tube or hypanthium, in which both sepals and petals are united in a short to long tube. In addition, the leaves are usually opposite and the perianth parts typically in fours with twice as many stamens. The diminutive, forest-floor genus *Circaea* is different in having a two-merous perianth. The aquatic genus *Ludwigia*, which lacks a floral tube, is exceptional in having persistent sepals and floral parts in (threes or) fours or fives (to sevens). The sepals of *Fuchsia* are also persistent, brightly colored, and often contrast with the petal color, contributing to floral display.

Selected Genera of Onagraceae
Camissonia • Chamaenerion • Circaea • Clarkia • Epilobium, including *Zauschneria • Fuchsia • Lopezia • Ludwigia • Oenothera,* including *Gaura*

Orchidaceae, *Sobralia macrantha*. Orchids are characterized by flowers with three outer and three inner tepals, both whorls petaloid, with one tepal in the inner whorl enlarged into a variously elaborated lip or labellum. The enlarged style is fused with the anther to form a thickened column or gynostemium. The inferior ovary matures into a capsule (rarely a berry) containing millions of dust-like seeds.

ORCHIDACEAE

Orchid Family

About 26,000 species in 760 genera

RANGE Cosmopolitan, especially tropics

PLANT FORM Perennial, mycotrophic (obtaining part or all of carbon, water, and nutrients through a symbiotic association with fungi), evergreen epiphytes, vines, some shrubs, or evergreen or deciduous terrestrial herbs, then with tubers, corms, or rhizomes, a few annuals, some that lack chlorophyll; stems of epiphytes sometimes forming pseudobulbs (bulb-like enlargements of the stem); leaves usually entire, in spirals or two-ranked, some in whorls, often fleshy, rarely scale-like

FLOWERS Highly specialized, usually bisexual, in racemes, panicles, or solitary, zygomorphic, often resupinate; perianth petaloid, of 6 tepals in 2 whorls, lowermost different from others, forming a lip or labellum, sometimes spurred from labellum or from one or more other tepals; stamens mostly 1 (or 2, rarely 3), when 1, united with style, forming a column (gynostemium), anthers with longitudinal slits, pollen often in clusters (pollinia), each with a stalk (caudicle); ovary inferior, of 3 united carpels, mostly 1-celled, with enlarged stigmatic lobe

FRUIT Dry capsules with three or six longitudinal slits but closed at base and apex, with minute, dust-like seeds

The largest family of flowering plants and together with Asteraceae accounting for some 50,000 species, Orchidaceae are mostly evergreen or deciduous herbs, or occasionally shrub-like. Most orchids occur in the tropics, usually as epiphytes. Orchidaceae are also well represented in regions of Mediterranean climate and in temperate grasslands where they are mostly seasonal geophytes with rhizomes or tubers. The flowers (for example, those of *Sobralia macrantha*, bamboo orchid) are very specialized and often difficult to interpret. The vast majority of genera have flowers inverted, thus twisted though 180 degrees (resupinate); the apparent lower tepal forms a lip-like structure (labellum) that is variously elaborated, even slipper-shaped in *Cypripedium* and allies (slipper orchids), which also have two stamens. Most other orchids have a single stamen fused with the style in a columnar structure called a gynostemium. The dust-like seeds of orchids germinate only in the presence of an appropriate fungus.

A few genera and species are hardy or half-hardy and are grown in temperate gardens. Most prominent of these is the eastern Asian genus *Bletilla;* pink-, blue-, and yellow-flowered species of this genus of geophytes adorn many temperate and Mediterranean gardens. *Pleione*, also deciduous, is more difficult to grow, but its surprisingly large flowers are rewarding. In gardens in Mediterranean and warmer temperate climates, species of *Dactylorhiza*, *Epipactis*, and even *Cymbidium* thrive

Orchidaceae, *Ophrys*

Orchidaceae, *Disa uniflora*

Orchidaceae, *Orchis tridentata*

Orchidaceae, *Disa longicornu*

outdoors. Many more are native in such climates, among these the bee orchid *Ophrys*, the labellum of which is patterned to resemble a female bee or wasp, the potential pollinators.

Despite the plethora of species, few orchids are useful to humans, although many are admired for the pleasure offered by their attractive and very long lasting blooms. Ground tubers of some are cooked and consumed in tropical Africa. A few species offer flavorings, most significantly vanilla from ripe capsules of the genus *Vanilla*. The Neotropical *V. planifolia* is cultivated, especially in Mexico. *Vanilla* is also grown extensively in Madagascar, where its pollinators are absent and flowers must be hand pollinated, providing a valuable export crop.

Selected Genera of Orchidaceae*
Angraecum • *Bletilla* • *Bulbophyllum* • *Cattleya* • *Cymbidium* • *Cypripedium* • *Dactylorhiza* • *Dendrobium* • *Disa* • *Epidendrum* • *Epipactis* • *Habenaria* • *Oncidium*, including many hybrids • *Ophrys* • *Orchis* • *Paphiopedilum* • *Phalaenopsis* • *Pleione* • *Satyrium* • *Serapias* • *Vanda* • *Vanilla*

*A list or genera of orchids, even just those in cultivation, would consume pages and seems unnecessary here as there are numerous books dealing with various aspects of the family. We list a few genera commonly found in gardens, native in warm temperate and Mediterranean climates, or grown as houseplants.

Papaveraceae subfamily Papaveroideae, *Papaver nudicaule*. As in most Papaveraceae, the calyx and corolla are two-parted, with two rapidly deciduous sepals and four free petals surrounding numerous stamens. In the genus *Papaver* the superior ovary consists of united carpels topped by sessile stigmas, thus without a style.

PAPAVERACEAE

Poppy Family

About 750 species in 43 genera

RANGE Mainly northern temperate, also tropical African mountains, southern Africa, and Australia

PLANT FORM Mostly perennial herbs, some scrambling or trailing, and several annuals, rarely vines or somewhat shrub-like; leaves simple, in spirals or a basal rosette, simple but usually lobed or somewhat to finely dissected, with tendrils in some vines, without stipules; mainly spring or summer blooming, often with white or yellow latex

FLOWERS Perfect, radially symmetric or zygomorphic, in cymes or solitary, in fumarioid genera often nodding or pendent; calyx with sepals usually 2, soon falling or persistent (sometimes absent), free or partly united, green or variously colored, then sometimes with pouches or spurs; corolla with 4 or 6 free petals in 2 whorls (lacking in *Macleaya*) or partly united (fumarioid group); stamens many or 4 or 6; ovary superior, compound, of 2 or many united carpels, 1-celled, with 1 style (or style lacking in most *Papaver* species) but stigmas often many, sometimes united

FRUIT A capsule splitting longitudinally or through pores, or nut-like in *Fumaria*

Traditionally, Papaveraceae have been regarded as a family separate from Fumariaceae, but they are now treated as a single family, with one of the subfamilies being the Fumarioideae. Papaveraceae provide many subjects for temperate gardens, notably *Papaver*, true poppies, which are traditionally separated from poppy-like *Meconopsis* by having a sessile versus a stalked stigma. Otherwise, the species of *Meconopsis* often, but not always, have remarkable, large, blue flowers. Several *Papaver* and *Meconopsis* species are in cultivation, among which the Iceland poppy, *P. nudicaule*, and the perennial *P. orientale* group are prominent. The genera *Romneya* and *Dendromecon* are exceptional in Papaveraceae in their shrub-like habit. The California poppy, *Eschscholzia californica*, though poppy-like in its flowers, has highly dissected leaves and very different fruits, elongate capsules that recall those of some *Dicentra* and *Corydalis* species (these belonging to the fumarioid group).

Bleeding heart (*Lamprocapnos spectabilis*, previously *Dicentra spectabilis*) is one of the most striking members of the poppy family. Among the many species of *Corydalis*, several including *C. elata* and *C. flexuosa* have unusual sky-blue flowers and are very fine garden plants. Other *Corydalis* species have considerable horticultural merit: *C. solida*, an early spring perennial, has bulb-like tubers and becomes dormant soon after blooming. Other fumarioids in cultivation include *Fumaria*, of which *F. capreolata* and *F. muralis* (wall fumitory) are often weedy. *Dactylocapnos scandens* (formerly a *Dicentra*) is a perennial vine with clusters of

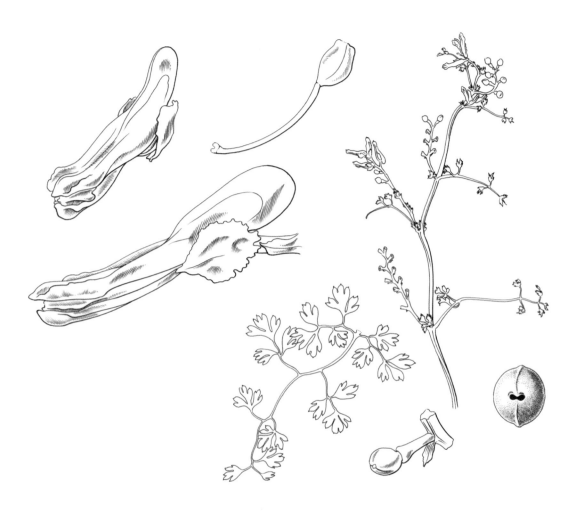

Papaveraceae subfamily Fumarioideae, *Fumaria muralis*. Fumitory and related genera are recognizable by their highly dissected leaves and compressed flowers with two small, scale-like sepals and four unequal petals, the two lateral petals joined at the tips and the uppermost pouched or spurred at the base. The six stamens are joined in two bundles, and the fruit in many species is a small nutlet.

Papaveraceae subfamily Papaveroideae, *Papaver*

Papaveraceae subfamily Papaveroideae, *Romneya coulteri*

Papaveraceae subfamily Fumarioideae, *Lamprocapnos spectabilis*

Papaveraceae subfamily Fumarioideae, *Corydalis triternata*

Papaveraceae subfamily Papaveroideae, *Papaver*

Papaveraceae subfamily Papaveroideae, *Meconopsis aculeata*

pendent yellow flowers and dissected leaves bearing tendrils. Several fumarioid genera occur in southern Africa, so far not in cultivation.

Flowering in woodland in early spring before leaves of deciduous trees have unfolded, *Eomecon* (snow poppy) and *Sanguinaria* (bloodroot) are sometimes grown in shady gardens. Another woodland poppy, *Sanguinaria*, has remarkably fern-like leaves, resembling those of sword fern or *Blechnum* species. Species of another woodland poppy, *Stylophorum*, have yellow poppy-like flowers produced well into late spring and summer.

Papaveraceae are recognizable primarily by their calyx, usually consisting of two sepals (three in *Romneya*), these soon falling in many genera, and a corolla of four free or partly united petals. The lobed to dissected leaves with milky or yellow latex are sometimes diagnostic. The fumarioid genera are recognizable by their highly dissected leaves and sometimes compressed flowers, with both sepals and petals colored but in separate whorls and of different appearance, the sepals sometimes with pouches or, in *Corydalis*, one petal bearing a long spur.

Selected Genera of Papaveraceae subfamily Papveroideae
Argemone • Chelidonium • Dendromecon • Eomecon • Eschscholzia • Glaucium • Macleaya • Papaver • Pteridophyllum • Roemeria • Romneya • Sanguinaria • Stylophorum

Selected Genera of Papaveraceae subfamily Fumarioideae
Adlumia • Corydalis • Cysticapnos • Dactylocapnos • Dicentra • Fumaria • Hypecoum • Lamprocapnos • Sarcocapnos

Passifloraceae, *Passiflora edulis.* The bizarre-looking flowers of passion flower have five free sepals and petals inserted on a hypanthium, and five stamens radiating from the top of a common stalk. The filamentous threads forming the corona also issue from the rim of the hypanthium. The superior ovary has three forked styles terminating in large stigmas.

PASSIFLORACEAE

Passion Flower Family

About 920 species in 29 genera

RANGE Tropical and warm temperate, especially South America

PLANT FORM Trees, shrubs, and evergreen vines, a few perennials; leaves in spirals, simple, entire or palmately lobed, rarely compound, usually without stipules; vines with axillary tendrils

FLOWERS Usually bisexual, in cymes or solitary in axils, with flat or tubular hypanthium (of united calyx and corolla tissue), often with stamens and style borne on a common stalk (androgynophore); calyx of 5–8 sepals, sometimes united basally and persistent; corolla of as many petals as sepals, often with a corona of scales or hairs on the rim of the hypanthium; stamens usually 5 (or many), sharing a common stalk with style, anthers with longitudinal slits; ovary superior, usually of 3 united carpels, 1-celled, with 1 style but 3-branched distally

FRUIT A leathery walled structure, technically a berry, containing many hard seeds enclosed in fleshy tissue, or a capsule

Passifloraceae used to be a small family of just a few genera, the only one of any size, *Passiflora*, with about 450 species. DNA studies show that the two tropical families Malesherbiaceae and Turneraceae should be united with Passifloraceae, significantly increasing its size. The family remains of limited importance in temperate horticulture. Few *Passiflora* species are frost tolerant, although several can be grown in areas of Mediterranean climate. The common name passion flower, first applied specifically to the Brazilian species *P. mucronata*, alludes to details of the crucifixion of Christ, the three styles recalling the three nails, five stamens the five wounds, and the corona, the crown of thorns.

Among hardy species of *Passiflora*, *P. incarnata* has large blue and green flowers, and *P. lutea* has smaller white to pale yellow flowers. *Passiflora coccinea* and *P. minata*, with bright red flowers, are perhaps the most striking of those available for gardens, but neither species is hardy.

Fruits of several *Passiflora* species are edible, best known being the granadilla or passion fruit, *P. edulis*. Its large globe-like fruits contain seeds with fleshy yellow tissue (an aril) with delicious, intense flavor. Too rarely seen in markets, and then expensive, the fruit is readily transported and keeps well yet has remained poorly known, at least in North America. Other species, including *P. ligularis* and *P. quadrangularis*, have similar fruit and are sold in markets in the tropics. Even the fruit of the eastern North American *P. incarnata* is edible and was consumed by native North American Indians. The species is fairly hardy, even in the lower Midwest, and its large blue and white flowers of unusual structure recommend it for gardens.

Selected Genera of Passifloraceae
Adenia • Malesherbia • Passiflora • Piriqueta • Turnera

Plumbaginaceae, *Plumbago auriculata*. Pale blue flowers are typical of Plumbaginaceae as is the calyx of united sepals and five-parted corolla with petals united below in a tube but with spreading limbs furled like an umbrella in bud. The five stamens scarcely protrude from the mouth of the tube.

PLUMBAGINACEAE

Plumbago Family

About 710 species in 30 genera

RANGE Cosmopolitan, especially maritime

PLANT FORM Shrubs and perennial herbs; leaves simple, entire or lobed to much dissected, in spirals or a basal rosette, without stipules, sometimes with chalk glands exuding calcium salts

FLOWERS Bisexual, radially symmetric, borne in panicles, cymose heads, or racemes; calyx of 5 sepals united to form a 5- or 10-ribbed tube, green or petaloid, lobes sometimes unequal; corolla forming a tube with 5 spreading lobes; stamens 5, opposite the corolla lobes, often not or barely emerging, anthers with longitudinal slits; ovary superior, of 5 united carpels, 1-celled, with separate styles or 1 style lobed at tip, with 1 basal ovule

FRUIT Dry, either an achene or capsule

Sometimes called the leadwort family, Plumbaginaceae are a relatively small family, contributing few species to temperate gardens. The southern African *Plumbago auriculata* (also known as *P. capensis*), a scrambling, softly woody shrub with pale blue flowers, is often seen in Mediterranean gardens and favors a warm temperate climate with no more than light frost. The relatively hardy *Ceratostigma*, commonly used as a ground cover, has attractive, deep blue flowers. More commonly grown is statice or sea lavender, now assigned to the genus *Limonium*, which has about 350 species, many of them maritime or growing in salt marshes. The genus is distinguishable by its papery, persistent calyx, colored blue or yellow in *L. sinuatum*, whereas the corolla is white. The inflorescences are used in dry floral arrangements as everlastings. The species has an unfortunate tendency to be invasive in dry areas, as in western South Africa. *Armeria* species, often called thrift or sea pink, are also occasionally grown for their tufted habit and crowded heads of small, pink or red blooms.

Plumbaginaceae, *Plumbago maritima*

Plumbaginaceae, *Plumbago auriculata*

Plumbaginaceae are recognizable by their simple leaves, inserted spirally (or in a basal rosette), five-merous flowers with a tubular or funnel-shaped calyx, tubular corolla with spreading lobes, and a superior, one-celled ovary with a single basal ovule.

Selected Genera of Plumbaginaceae
Acantholimon • *Armeria* • *Ceratostigma* • *Limonium*, including *Statice* • *Plumbago*

POACEAE

Grass Family

About 11,300 species in more than 750 genera

RANGE Cosmopolitan, especially northern temperate and dry tropics

PLANT FORM Mostly perennial herbs, some annuals, also tree-like (bamboos), either grow-ing in tufts or spreading, stems round in cross section, often with hollow internodes and thickened nodes; leaves in two ranks, spirals, or a basal rosette, simple, nearly always with parallel venation, with a sheathing base and spreading blade, usually with an outgrowth (ligule) at the base of the blade, without stipules

FLOWERS Highly modified from the typical, thus with specialized terminology, perfect or unisexual, sessile and clustered in so-called spikelets and enclosed by scale-like bracts, often with awns; spikelets borne in small to large compound inflorescences; flowers without calyx or corolla (these represented by minute scales called lodicules); stamens usually 3 (or 2 or 1, in bamboos sometimes many), free; ovary of 3, 2, or 1 carpel(s), with 1 ovule, with 2 free styles and feathery

FRUIT An achene, so a single seed enclosed by a thin layer derived from the outer wall of the ovary, thus grain-like, small or large, and shed as a unit

Perhaps traditionally not regarded as worth horticultural attention except as a lawn, today many species of grasses (Poaceae, historical name Gramineae) are grown in gardens for display as sculptural elements. Some can be dramatic in flower or fruit, with feathery inflorescences that remain impressive long after flowering, maintaining their distinctive form in seed. Other species have colored foliage, for example, Japanese blood grass (a cultivar of *Imperata cylindrica*) and Japanese forest grass (*Hakonechloa*), cultivars of which have gold- or red-streaked leaves or uniformly yellow-green foliage. Cultivars of *Miscanthus chinensis* bearing leaves with transverse or longitudinal pale streaks (zebra grass) are grown not only for their foliage but the handsome inflorescences that persist into the winter months. Northern sea grass (*Chasmanthium latifolium*) is grown for its attractive seed heads with large overlapping bracts, sometimes used in dry floral arrange-ments. A variegated cultivar is especially attractive.

Several grasses are used in lawns, including *Cynodon dactylon* (Bermuda or kweek grass), *Festuca* (bluegrass), and *Zoysia*. The South American genus *Cortaderia* (giant or dwarf pampas grass) is grown for the handsome, plumose inflorescences. Sadly, pampas grass is becoming invasive and hard to control in New Zealand, western North America, southern Africa, and elsewhere. Several bamboos (subfamily Bambusoideae) are grown for ornament, their distinctive leaves and stems providing unique structural elements to any garden. Some are grown for their unusually colored stems, sometimes black, golden, or striped, or their variegated foliage. Among the more important genera are the clump-forming *Fargesia* and the spreading *Phyllostachys*.

In contrast to their relatively modest horticultural importance, Poaceae are the primary plant family for providing food for humans, including barley (*Hordeum*), corn (maize, *Zea mays*), rice (*Oryza*), rye (*Secale*), and wheat (*Triticum*), to mention only the most important. *Triticum aestivum*, the common bread wheat, has a complex, incompletely understood genetic history with an ancestry involving hybridization with species of other grass genera. Likewise, the ancestry of corn is uncertain; it was certainly developed in Mesoamerica by deliberate or unconscious selection from wild ancestors that are now classified as different genera. No plant resembling corn ever existed in the wild, and it thus represents a striking example of a genetically modified organism. More cropland is devoted to these grains than to any other food. Grasses also provide the bulk of food eaten by cattle, goats, and sheep. Some species of bamboo grow to 35 feet (10 meters) high or more and yield timber for construction, furniture, and utensils. As most gardeners know, once you have bamboo, it is hard to eradicate and can be invasive, spreading by underground rhizomes rather than seeds.

Poaceae are distinguishable from similar graminoid (grass-like) families by their hollow stems, round in cross section and with thickened nodes, leaves with a sheathing base wrapped around the stem, usually narrow leaf blades with parallel venation, and the ligule, a small scale- or collar-like outgrowth present at the base of the blade. Details of the inflorescence, spikelets, and flowers are needed to recognize the subfamilies, tribes, and genera of the family. The so-called reeds or sedges (Cyperaceae), related to the grasses, mostly have stems triangular in cross section and pith-filled. The rushes (Juncaceae), not included in this book, may be confused with grasses but can be distinguished by leaves crowded basally, stems round in cross section, and flowers of conventional structure with small though recognizable perianth parts. Some members of the grass-like Restionaceae, also not included here, or restios,

are receiving increasing horticultural attention and are recognizable by having leaves reduced to dry scales; the green "leafy" organs are actually branches and thus are round in cross section, without the expanded blades expected in true leaves.

Selected Genera of Poaceae

Aegilops • Chasmanthium • Cortaderia • Cynodon • Eragrostis • Fargesia • Festuca • Hakonechloa • Hordeum • Imperata • Miscanthus • Oryza • Panicum • Pennisetum • Phyllostachys • Pseudosasa • Sasa • Secale • Triticum • Zea • Zizania • Zoysia

POLEMONIACEAE

Phlox Family

About 375 species in 22 genera

RANGE Eurasia and North and South America, especially western North America, mostly temperate climates

PLANT FORM Small trees, shrubs, vines, and many perennial herbs, a few annuals; leaves usually simple, entire to pinnately lobed or rarely compound, usually in spirals, opposite (*Phlox*), or rarely whorled, without stipules

FLOWERS Bisexual, in head-like cymes, racemes, or solitary, radially symmetric or zygomorphic, mostly 5-merous or 4- or 6-merous; calyx with sepals often united for some distance; corolla with petals united in a wide to narrow tube, as many as sepals; stamens as many as sepals, inserted on corolla tube, anthers with longitudinal slits; ovary superior, compound, of (2 or) 3 (or 4) united carpels with as many cells, style terminal with as many lobes as carpels

FRUIT A capsule, sometimes splitting explosively, usually with many small seeds

Polemoniaceae, *Polemonium confertum*

Polemoniaceae, *Phlox longifolia*

Few temperate gardens are without a species of two of the largest genus of Polemoniaceae, *Phlox*, which has some 70 species. The common perennial phlox of gardens is *P. paniculata*, of which there are many cultivars, the flowers ranging from pink or white to purple and nearly red. The narrow-leaved *P. subulata* is a mat-forming perennial used as a ground cover or in rock gardens. The annual *P. drummondii* is also widely cultivated, and several more species are available in the nursery trade.

Blue- or pink-flowered *Polemonium caeruleum* (Jacob's ladder) is sometimes encountered in gardens, and a cultivar with purplish red foliage is especially attractive. Red-flowered species of *Gilia* and *Ipomopsis* are also grown in gardens. The cup and saucer vine, *Cobaea scandens*, has unusually large flowers for the family. Of tropical origin, is must be treated as an annual in temperate climates.

Polemoniaceae are recognizable by their tubular calyx (not in some tropical species) with as many lobes as the corolla (usually five), a tubular corolla, usually five stamens, and leaves in spirals except in *Phlox*, in which they are opposite.

Selected Genera of Polemoniaceae
Acanthogilia • Cantua • Cobaea • Collomia • Gilia • Ipomopsis • Linanthus • Loeselia • Navarretia • Phlox • Polemonium

POLYGALACEAE

Milkwort Family

About 700 species in 24 genera

RANGE Nearly cosmopolitan, tropical and temperate

PLANT FORM Some trees, vines, and many shrubs, some herbs; leaves simple, usually in spirals, sometimes spiny or ericoid, entire, mostly without stipules

FLOWERS Bisexual, subtended by 2 small bracts, zygomorphic, in racemes, spikes, or panicles, mostly 5-merous; calyx of 5 sepals sometimes united basally, the 2 inner often petaloid; corolla mostly of 3 (to 5) petals, lower median often boat-shaped and fringed at tip; stamens 8 or 10, united basally or in a tube slit dorsally, anthers mostly with longitudinal slits, rarely apical pores; ovary superior, compound, of 2–5 (to 8) united carpels, mostly 2- to 5-celled, with terminal style often curved and 2-lobed, 1 lobe ending in a tuft of hairs

FRUIT A capsule, nut, drupe, or winged (a samara)

Polygalaceae, *Polygala meridionalis*

By far the best-known genus of Polygalaceae is *Polygala* with about 200 species, several of which are grown in gardens. The southern African *P. myrtifolia* (butterfly bush) and *P. virgata* are often seen in southern-hemisphere gardens, less often in those in the northern hemisphere with Mediterranean or warm temperate climates. Both have become invasive in Australia. The shrub or alpine *P. chamaebuxus* (bastard box) is grown for its striking flowers with yellow corolla and purple wings. Several more *Polygala* species are grown as alpines. The African genus *Muraltia* consists mostly of small shrubs, and some are cultivated in South Africa and perhaps elsewhere, especially *M. heisteria*, which has relatively large purple flowers and leaves with sharp tips.

The mostly arborescent genus *Securidaca* is largely tropical, and several species are grown but are unknown in temperate and even Mediterranean gardens. The genus includes several with medicinal properties. The North American *Polygala senega* (seneca snakeroot) has been used as a cathartic or emetic, and some members of the genus are reputedly antidotes for snakebite. The generic name *Polygala* is derived from the Greek (*poly*, much, and *gala*, milk) because plants supposedly stimulated lactation in cattle. This is evidently the source of the common name milkwort for the genus and family. Species of Polygalaceae do not have milky latex as the common name might suggest.

Species of Polygalaceae are often confused with Fabaceae (pea family), especially those with flowers like pea and bean plants; careful examination shows the fringe at the tip of the larger, boat-shaped, lowermost petal that is characteristic of Polygalaceae but not present in Fabaceae. Leaves of Polygalaceae are simple, sometimes needle-like, unlike the compound leaves of most (but not all) members of the pea family.

Selected Genera of Polygalaceae
Monnina • Muraltia • Polygala • Securidaca • Xanthophyllum

Polygalaceae, *Polygala myrtifolia.* All Polygalaceae have simple, spirally inserted leaves, and distinctive flowers that superficially resemble those of papilionoid legumes. The flowers of *Polygala* have five unequal sepals, the upper two laterals greatly enlarged and brightly colored, and three very unequal petals, the upper two reduced but the lower median one forming a keel-like structure bearing a brush-like crest at the tip and enclosing the stamens and style. The eight stamens are joined into an open tube, and the anthers shed their pollen through apical pores directly onto the sterile upper lobe of the stigma, which dabs it onto visiting insects that are heavy enough to depress the hinged keel.

Polygalaceae, *Polygala myrtifolia*

POLYGONACEAE

Buckwheat Family

About 1200 species in 52 genera

RANGE More or less cosmopolitan, especially northern temperate

PLANT FORM Some trees, shrubs, vines, many perennial herbs, a few annuals; stems usually jointed with swollen nodes; leaves simple, usually in spirals or sometimes clustered basally (at least early in season), usually with prominent stipules forming a dry, papery or green sheath around stem (lacking in *Eriogonum*)

FLOWERS Usually bisexual, radially symmetric, in racemes or spikes, or umbels or heads in *Eriogonum;* mostly 5-merous (or 2- or 3-merous), not differentiated into calyx and corolla, thus perianth forming a minute tube and lobes spreading, sometimes green, more often variously colored; stamens as many as tepals or as many as 8, inserted at base of ovary, anthers with longitudinal slits; ovary superior, compound, of (2 or) 3 (or 4) united carpels but 1-celled with 1 basal ovule

FRUIT Usually a three-lobed achene or nutlet, sometimes enclosed by a persistent perianth, this sometimes forming a wing

Polygonaceae contribute only modestly to ornamental horticulture and likewise to our range of edible plants. For the garden, *Persicaria* provides some species with striking leaf patterning, as in *P. virginianum*, the flowers of which are minute, whereas *P. amplexicaulis* has larger red flowers produced in profusion in crowded racemes in mid to late summer. *Persicaria bistorta* (bistort) has pink or white flowers borne in attractive, crowded spikes. Some species of *Rumex*, widely known as dock, are also grown for their patterned leaves. *Antigonon leptopus*, a beautiful twining species, sometimes called coral vine, has pink or white flowers. *Muehlenbeckia complexa* (also known as *M. axillaris*) is another attractive vine, native to New Zealand, with small leaves and even smaller flowers, sometimes called maidenhair vine. *Eriogonum*, one of the largest plant genera in North America, includes several species worth the attention of gardeners. *Eriogonum umbellatum* has bright yellow flowers in tight umbels, sometimes with the buds flushed orange; many more species of this genus are cultivated locally, especially in gardens devoted to native flora.

The genus *Rheum* includes edible rhubarbs, the petioles of which are eaten after cooking, and several more ornamental species, some with magnificent leaves or large inflorescences such as the giant herb *R. palmatum*, up to 6 feet (2 meters) tall. Polygonaceae also include *Fagopyrum* (buckwheat), the seeds of which are eaten as a grain or ground into flour. Perhaps more commonly eaten in the past, buckwheat pancakes are a fine alternative to those made from wheat flour. Leaves of *Rumex acetosa* (sorrel), used to flavor soups and sauces, have a tangy, slightly sour, refreshing taste. *Fallopia japonica* (knotweed) was introduced from eastern Asia as an ornamental and has become an invasive weed, expensive to eradicate. Other weedy species are *Emex australis* and *E. spinosa*, found in pastures in dry country; both have large spiny fruits.

Most Polygonaceae are recognizable by their stems with swollen nodes and the leaf stipules forming a sheath around the stem (an ochrea), usually dry and brown-papery but sometimes green. *Eriogonum* species lack an ochrea, making family identification difficult, but the nodes are slightly woolly or downy. The three-lobed fruits of many Polygonaceae aid in identification when present.

Selected Genera of Polygonaceae
Antigonon • Emex • Eriogonum • Fagopyrum • Fallopia • Muehlenbeckia • Persicaria • Polygonum • Rheum • Rumex

Polygonaceae, *Persicaria amplexicaulis*

Polygonaceae, *Eriogonum umbellatum*

PORTULACACEAE

Purslane Family

About 325 species in 20 genera

RANGE Nearly cosmopolitan, especially North and South America

PLANT FORM Shrubs and herbs, including annuals, often succulent; leaves simple, in spirals or opposite, sometimes clustered basally, stipules dry or tufts of hairs or lacking

FLOWERS Bisexual, usually radially symmetric, solitary, in cymes, or in heads; calyx of 2 or 3 sepals or as many as 9 in *Lewisia*, usually succulent and persistent; corolla with petals partly united, (2- or) 5- to 18-lobed, usually free; stamens as many as and opposite petals or many in bundles, anthers with longitudinal slits; ovary superior or inferior, compound, of 2–9 united carpels, eventually 1-celled with 1 to many ovules, with 1 terminal style or lobed style

FRUIT Usually a capsule, sometimes with circular dehiscence (circumscissile)

Portulacaceae are closely related to Cactaceae but are typically small succulent herbs or shrubs without spines. *Portulaca oleracea* (purslane), a cosmopolitan weed, is used as a potherb and is rich in vitamins. The portulaca of gardens, *P. grandiflora*, is a sprawling annual with ephemeral yellow, red, or orange flowers. Species of *Calandrinia* have striking, large, pink flowers and are occasionally grown in gardens. Leaves of some *Claytonia* and *Montia* species, known as spring beauty or miner's lettuce, make an interesting addition to salads and are sometimes grown in gardens, especially *C. perfoliata*. The most attractive Portulacaceae are species of the western North American genus *Lewisia*. Unlike most of the family, the long-lasting flowers of *Lewisia* species have numerous (as many as 18) petals and are fine garden subjects, especially when grown as alpines with good drainage. *Lewisia* (sometimes called bitterroot) is the state flower of Montana, and its common name is celebrated in the Bitterroot Mountains of that state. Species of

succulent *Anacapseros* and *Talinum* are occasionally grown in specialist collections of dwarf succulents.

The family is recognizable by its succulent but not spiny leaves and stems, mostly two sepals and ephemeral petals, and the one-celled ovary with many ovules. The exception is *Lewisia*, which has long-lasting flowers with as many as 9 sepals and 18 petals. An alternative treatment of Portulacaceae renders the family with just one genus, *Portulaca; Calandrinia, Claytonia, Lewisia,* and *Montia* are removed to Montiaceae, *Anacampseros* to Anacampserotaceae, and *Talinum* to Talinaceae.

Selected Genera of Portulacaceae
Anacampseros • Calandrinia • Claytonia • Lewisia • Montia • Portulaca • Talinum

Portulacaceae, *Portulaca trianthemoides*

Portulacaceae, *Lewisia cotyledon*

PRIMULACEAE

Primrose Family

About 2645 species in 57 genera

RANGE Nearly cosmopolitan, especially tropics and northern temperate zone

PLANT FORM Evergreen trees, shrubs, vines, and herbs, including many perennials; leaves simple, often toothed, in spirals or whorls and often clustered basally, without stipules

FLOWERS Bisexual, usually radially symmetric, mostly 5-merous, often in umbels or heads, sometimes solitary (*Cyclamen*), or in racemes, panicles, or pagoda-like whorls; calyx of 5 basally united sepals (3 in *Anagallis*), often persistent; corolla with petals partly united, (3- or) 5-lobed; stamens as many as petals, inserted on the corolla tube; ovary compound, of (3 or) 5 united carpels but 1-celled, superior (rarely half-inferior), with 1 terminal style, ovary with 5 to many ovules

FRUIT Usually dry capsules, often splitting in a horizontal ring (circumscissile), rarely a fleshy drupe

Primulaceae, now including genera of the former family Myrsinaceae (but few of these in cultivation), are best known for their beloved hardy perennials, most importantly *Primula* (primrose, cowslip) with more than 400 species, but also the related genera *Cortusa*, *Dionysia*, and *Dodecatheon* (shooting star), these sometimes included in *Primula*. Other cultivated genera are *Cyclamen*, which has a bulb-like rootstock, and the morphologically heterogeneous *Lysimachia*, which includes creeping, herbaceous and shrubby species (creeping jenny, yellow loosestrife, gooseneck loosestrife). *Anagallis*, with the annual and often weedy *A. arvensis* (common or scarlet pimpernel), belongs in the family and contains extracts useful in treating polio and herpes. Several species of the genus *Androsace* (including the former *Douglasia*) are attractive perennials for the alpine and rock garden. The herbaceous *Trientalis*, with one species in Europe and one in North America, is occasionally seen in cultivation. Primulaceae also include many woody tropical trees and shrubs, few even known by name to gardeners.

The family can often be recognized by the umbellate inflorescence (or flowers in pagoda-like whorls with a terminal umbel) in *Dodecatheon* and *Primula*. Flowers have petals united basally or forming a narrow tube and spreading petal limbs. Most species of *Primula* are heterostylous, thus individuals may have flowers with either the anthers reaching the mouth of the tube and the style held within the tube, or anthers within the tube and style reaching the mouth. Differences in the anthers and style are associated with different pollen grains and stigma surfaces. The fruit of most Primulaceae is a capsule with distinctive circular dehiscence but is not often visible at flowering time.

Selected Genera of Primulaceae

Anagallis • *Androsace,* including *Douglasia* • *Ardisia* • *Clavija* • *Cortusa* • *Cyclamen* • *Dionysia* • *Discocalyx* • *Dodecatheon* • *Embelia* • *Glaux* • *Lysimachia* • *Myrsine* • *Oncostemon* • *Primula* • *Samolus* • *Soldanella* • *Theophrasta* • *Trientalis*

Primulaceae, *Cyclamen persicum*

Primulaceae, *Primula macrocalyx*

Primulaceae, *Primula japonica*

Primulaceae, *Androsace sempervivoides*

Proteaceae, *Stenocarpus sinuatus*. The bright red flowers are borne in an umbel, each flower consisting of four slender, colorful sepals joined at the base. The four sessile anthers are attached directly to the tips of the sepals. Pollen is shed onto the outside of the capitate stigma in bud, from where it is brushed onto visiting birds foraging for nectar secreted into the base of the floral tube.

PROTEACEAE

Protea Family

About 1710 species in 75 genera

RANGE A classic southern-hemisphere family, most diverse in Australasia but also well developed in South Africa, and few in tropical Africa, Madagascar, and South America

PLANT FORM Evergreen trees and shrubs, some prostrate but not herbaceous; leaves simple, usually in spirals, sometimes compound, often thick and leathery, without stipules

FLOWERS Usually bisexual, radially symmetric or zygomorphic, usually in racemes, umbels, or heads, sometimes surrounded by large colored bracts; calyx of 4 sepals, green or petaloid, partly united in a tube, this deeply split on 1 side, or 1 sepal free; corolla present as scales or glands, or lacking; stamens opposite calyx lobes and inserted on them, anthers with longitudinal slits, 1 theca often sterile; ovary superior, of 1 carpel with 1 or 2 to many ovules, style elongate, sometimes pollen shed onto style below apex, thus serving as a pollen presenter

FRUIT Usually dry achenes (one-seeded) or follicles (many-seeded), sometimes drupes or nuts

This very ancient family offers only a modest contribution to temperate horticulture, but many can be grown in a Mediterranean climate. Moreover, introduction of species from high elevations is likely to provide several interesting additions to temperate gardens. Most genera have relatively small flowers, but these are often aggregated in large inflorescences, thereby rendering them desirable garden plants. Currently, several species of the Australian genera *Banksia* and *Grevillea* are hardy in U.S. Department of Agriculture Plant Hardiness Zone 7. Some cultivars of the tree *Embothrium coccineum* (Chilean firebush) are also hardy and particularly attractive plants, having large inflorescences of red flowers. The Australian *Telopea speciosissima* (waratah) has beautiful, carmine-red, *Protea*-like inflorescences. Another small Australian tree, *Stenocarpus sinuatus* (firewheel tree), has pillar-box-red flowers, a striking plant for Mediterranean and warmer temperate gardens. Many genera and species of the family are grown in gardens in Australia, New Zealand, and South Africa.

Proteaceae are immediately recognizable by their prominent calyx of four sepals, these partly united into a tube split down one side, four stamens joined to the calyx, the corolla often reduced to scales or lacking, and an ovary of a single carpel with an elongate style. The flowers are often borne in racemes or heads, then sometimes subtended by large bracts. In *Protea*, the flower heads are surrounded by particularly large, colored bracts.

Some species of the family are grown for the florist trade, especially several of *Protea* as well as *Leucadendron*, the latter often for its colorful foliage, and *Leucospermum* (pincushion). The cultivar *Leucadendron laureolum* 'Safari Sunset' has dark red leaves and stems. *Macadamia* species and hybrids are grown in the tropics and some subtropical areas for their oil-rich and delicious nuts.

Selected Genera of Proteaceae
Aulax • *Banksia* • *Embothrium* • *Faurea* • *Grevillea* • *Hakea* • *Leucadendron* • *Leucospermum* • *Macadamia* • *Persoonia* • *Protea* • *Serruria* • *Stenocarpus* • *Telopea*

Proteaceae, *Mimetes cucullatus*

Proteaceae, *Protea dracomontana*

Proteaceae, *Leucospermum conocarpodendron*

Proteaceae, *Leucospermum reflexum*

Proteaceae, *Leucadendron salignum*

Ranunculaceae, *Anemone japonica.* Most Ranunculaceae are herbaceous, often with palmately lobed leaves, and many, such as anem-
ones, have a perianth of multiple petal-like sepals, numerous spirally
arranged stamens, and many crowded carpels that often mature into
individual, wind-dispersed fruits with feathery stigmas.

RANUNCULACEAE

Buttercup Family

About 2100 species in 55 genera

RANGE Mostly northern temperate, also tropical mountains

PLANT FORM Mostly herbs, some vines, a few small shrubs, and aquatics; leaves simple, often palmately lobed or much dissected to compound (some *Clematis*), in spirals, crowded basally, or opposite (*Clematis*), with minute or no stipules, often with broad bases

FLOWERS Usually bisexual, usually radially symmetric, sometimes zygomorphic (*Aconitum*, *Delphinium*), usually in simple or compound cymes, racemes, or solitary; calyx 3–8 or more free sepals, green or petaloid; corolla with free petals, few to many or lacking; stamens many in spiral arrangement, inserted on receptacle; ovary superior, of few to many free carpels (fused below in *Nigella*), each with a separate stigma and with 1 to many ovules, stigmas 2-lobed or capitate

FRUIT Usually dry achenes (one-seeded), sometimes with feathery styles, or follicles (many-seeded), sometimes berries, or a capsule (*Nigella*)

Few families offer more to the horticultural world than Ranunculaceae, and hardly a temperate garden anywhere is without several members of this family. *Anemone* (for example, *A. japonica*) and *Clematis* actually lack true petals, the sepals of the calyx constituting the colorful parts of the flowers. *Clematis* comprises a few shrubs but many vines, the leaves with twining leaf stalks, but many other species are upright perennials. *Anemone* includes corm-bearing perennials, notably the Mediterranean *A. coronaria* and the ephemeral woodland species *A. nemorosa* among others. The pasqueflower (*A. pulsatilla*, previously *Pulsatilla vulgaris*) is an attractive early-flowering species that has unusual, persistent, feathery stigmas showing long after the flowers themselves fade.

Molecular studies show that species of the African genus *Knowltonia*, as well as *Hepatica* and *Pulsatilla*, are nested in *Anemone* and belong in that genus. The hepatica group, named for the three-lobed leaves, recalling the lobes of a liver, are treasured for their early spring blooming although the flowers are quite modest in size. The genus *Ranunculus* (buttercups) is remarkably variable and includes some floating aquatics, the Eurasian *R. asiaticus* (the ranunculus of gardens), and many weedy or invasive species. *Caltha palustris* (marsh marigold) is another invasive member of the family. *Helleborus* and its close relative *Trollius* are winter- and spring-blooming perennials. Flowers of *Aconitum*, *Aquilegia*, and *Delphinium* are more complex in arrangement, with one or more petals bearing spurs containing nectar.

Ranunculaceae, *Aquilegia vulgaris*

Ranunculaceae, *Anemone nemorosa*

Ranunculaceae, *Anemone hepatica*

Ranunculaceae, *Caltha palustris*

Many Ranunculaceae contain toxic compounds, including benzylisoquinoline, various alkaloids, and cardiac glycosides. Species of *Actaea* (which now includes *Cimicifuga*), with some species known as bugbane, are poisonous and used medicinally. At least *A. foetida* was used in eastern Asia to deter bugs. *Aconitum* (wolfbane) contains alkaloids extremely toxic to animals, including humans, but is also used in medicine. Few members of the family produce edible products, but exceptions are *Adonis* and *Nigella sativa* (love-in-a-mist), the seeds of which are a food flavoring, often sprinkled on breads. Many more species of Ranunculaceae are used in traditional medicine across the world. *Hydrastis canadensis* (golden seal) is used as a multipurpose remedy believed to possess many different medicinal properties. In addition to being used as a topical antimicrobial, it is taken internally as a digestive aid. Golden seal is available as a salve, tablet, or tincture and may be used to boost the medicinal effects of other herbs.

Selected Genera of Ranunculaceae

Aconitum • *Actaea,* including *Cimicifuga* • *Adonis* • *Anemone,* including *Hepatica, Knowltonia,* and *Pulsatilla* • *Anemonopsis* • *Aquilegia* • *Caltha* • *Clematis* • *Coptis* • *Delphinium* • *Eranthis* • *Helleborus* • *Isopyrum* • *Hydrastis* • *Nigella* • *Paraquilegia* • *Ranunculus* • *Thalictrum* • *Trollius*

PAEONIACEAE The genus *Paeonia*, which includes several species grown in temperate gardens, was long thought to be a member of Ranunculaceae, but newer information shows it to be more closely related to Saxifagaceae and best treated as a separate family. The genus has several features also found in Ranunculaceae, including multiple petals in spiral arrangement, multiple stamens, and free carpels, but the large hard-coated seeds are unlike those of any Ranunculaceae. Peonies include both woody shrubs and herbaceous perennials with simple, lobed, to much-dissected leaves. The family is restricted to Eurasia and western North America, entirely in the northern temperate zone.

RHAMNACEAE

Buckthorn Family

About 1050 species in 63 genera

RANGE Cosmopolitan, especially tropics and warm temperate, also western North America

PLANT FORM Trees, shrubs, often thorny, some vines, rarely herbaceous perennials; leaves simple, in spirals or opposite, always entire, often toothed, sometimes small, occasionally needle-like, usually with small, sometimes spiny stipules or these lacking, often with three to five main veins from base of blade

FLOWERS Usually small, often unisexual, radially symmetric, in racemes, spikes, cymes, fascicles (or solitary), 4- or 5-merous, with cupped hypanthium or hypanthium fused to inferior ovary; calyx mostly of 4 or 5 free sepals; corolla of free petals as many as sepals or lacking, petals sometimes hooded over stamens or twisted through 90 degrees; stamens as many as petals, alternating with sepals, anthers minute, splitting longitudinally; ovary superior or inferior, of 2-4 united carpels with as many cells, usually each with 1 ovule, with terminal style divided distally

FRUIT Usually a fleshy, one- or several-seeded drupe, or a dry capsule, or separating into separate sections, sometimes winged

A family of modest size, Rhamnaceae provide few species for temperate gardens. A form of *Rhamnus alaternus* (Italian or Mediterranean buckthorn) with variegated leaves is an attractive small tree and often used as hedging or topiary. Purple-flowered *R. prinoides* is occasionally grown as a small evergreen tree for gardens in southern Africa and Australia. Rhamnaceae include two large genera, most importantly the North American and mostly western *Ceanothus* (55 species) (California lilac) and the African and Atlantic *Phylica*. Several species of *Ceanothus* are hardy and widely cultivated for their blue flowers, and numerous selections and cultivars are available in the nursery trade, including as ground covers, some with variegated leaves. *Phylica*, with some 188 species, includes a few species cultivated for their large, feathery, and persistent bracts subtending the inconspicuous flowers. Many more *Ceanothus* and *Phylica* species are grown in gardens devoted to native flora.

Rhamnaceae include the genus *Ziziphus* (often seen spelled *Zizyphus*), which are mostly evergreen trees of the dry tropics, Mediterranean, and Middle East. Some species are used as street trees, and several as timber for their hardwood. The fleshy fruits of *Z. jujuba* (the jujube) and *Z. lotus* (lotus fruits of Greek mythology) are eaten fresh or dried or made into candy. *Ziziphus spina-christae* (Christ thorn) is possibly the plant used for Christ's crown of thorns, and *Z. mauritiana* (Indian jujube) is an invasive and noxious weed in Australia. *Rhamnus cathartica* berries and bark provide a purgative medication.

Rhamnaceae are recognizable by their spirally arranged leaves, either with toothed margins and often three to five main veins from the base of the blade, or small and ericoid, notably in *Phylica*, and flowers with a hypanthium on which the free sepals and petals (when present) are attached. The stamens are inserted alternate to the sepals, thus opposite the petals.

Selected Genera of Rhamnaceae

Ceanothus • Colletia • Colubrina • Frangula • Helinus • Paliurus • Phylica • Pomaderris • Rhamnus • Scutia • Trichocephalus • Ventilago • Ziziphus

Rhamnaceae, *Phylica plumosa*

Rhamnaceae, *Ceanothus* hybrid

Rosaceae, *Rosa chinensis* 'Mutabilis'. Typical of the rosoid group of the family, wild roses have five free sepals and petals, numerous stamens, and several free carpels, seemingly inferior but actually partly enclosed by a cupped receptacle. Each carpel bears a long style that extends outside the receptacle. In *Rosa* the paired stipules are joined to the base of the petiole. In fruit the receptacle becomes somewhat fleshy, called a hip, and encloses the ripe seeds.

ROSACEAE

Rose Family

About 3000 species in 85 genera

RANGE Almost cosmopolitan, especially temperate and warm northern hemisphere

PLANT FORM Moderate-sized trees, shrubs, and subshrubs, some vines, many perennials; leaves mostly in spirals (rarely opposite, for example, *Rhodotypos*), often compound, or simple then sometimes dissected, usually with stipules (not *Spiraea*), these sometimes conspicuous and joined to petiole; many deciduous, sometimes evergreen (for example, some *Prunus* species, and *Eriobotrya*, the loquat)

FLOWERS Usually perfect and radially symmetric (zygomorphic in *Gillenia*) or unisexual, sometimes without petals, terminal in cymes or in stalked clusters in the axils of leaves; calyx usually of 5 (to 10) free sepals, sometimes borne on hypanthium, green; corolla with 5 (to 10) free petals, often large; stamens usually many in sets of 5, inserted on receptacle; ovary superior, of 1 (prunoid) to many free or united carpels, with separate styles and stigmas

FRUIT A head of follicles or achenes (sometimes on a fleshy receptacle), sometimes enclosed in an enlarged receptacle (rose hips), or heads of fleshy, one-seeded drupes ("berries" as in raspberry), or solitary drupes or nuts

An important family for horticulture and the cut-flower market, Rosaceae were traditionally subdivided into four subfamilies: rosoids (Rosoideae), maloids (Maloideae), prunoids (Prunoideae), and spiraeoids (Spiraeoideae), each with very different fruits. That classification is now superseded by one in which 17 tribes are recognized. Nevertheless, the old classification remains useful in discussing the family. *Aronia*, a maloid, provides fruit called chokeberries.

Spiraeoids have small, often white, sometimes pink flowers, multiple small free carpels, and dry fruits (one-seeded, then achenes, or many-seeded, then follicles). Prunoids have larger flowers and drupaceous fruits with a single seed enclosed by a woody inner layer and a fleshy to leathery outer layer, thus including *Prunus* (apricots, cherries, plums, peaches, and even almonds). Maloids have small to large fleshy fruits in which the fleshy part is derived from the floral receptacle that enlarges during development to enclose the compound, several-seeded ovary, thus including *Malus* (apples), *Pyrus* (pears), and *Cydonia* (quinces). Rosoids have a leathery fruit, called hips. *Rubus* species, the brambles, have heads of fleshy, small, one-seeded drupes (for example, blackberries and raspberries).

Spiraeoids in cultivation include *Alchemilla* (including lady's mantle), *Aruncus* (goat's beard), *Filipendula* (burnet, queen of the prairie), *Holodiscus* (ocean spray), *Lyonothamnus*, *Physocarpus* (nine bark), and *Spiraea*. Genera of this group are sometimes conflated with *Astilbe*, which although broadly similar in

Rosaceae, *Geum elatum*

Rosaceae, *Geum capense*

Rosaceae, *Rubus bergii*

Rosaceae, *Spiraea thunbergii*

Rosaceae, *Aronia arbutifolia*

Rosaceae, *Rosa*

Rosaceae, *Pyrus syriaca*

flower is a genus of Saxifragaceae and distinguishable by the absence of stipules and an ovary of two partly united carpels, as well as having highly dissected leaves. Such leaves are rare in Rosaceae although characteristic of goat's beard and some *Filipendula* species (but the leaves in these genera have stipules). Occasionally cultivated, the North American shrub-like woody perennial *Gillenia* has zygomorphic flowers, unusual for Rosaceae, but the follicular fruits characteristic of spiraeoids.

Few temperate gardens are without roses (*Rosa*), many of the cultivated forms being hybrids between two or more wild species, or sometimes selections of a species (for example, *R. chinensis* 'Mutabilis', China rose or butterfly rose). The compound leaves have distinctive stipules fused to the bases of the petioles,

and the wild species have characteristic leathery fruits called hips. Standing out in the family for their opposite leaves are the genera *Lyonothamnus* and *Rhodotypos*. The latter has flowers like *Rubus* but dry fruits and large black seeds. Many more Rosaceae may be found in gardens, notably *Geum*, a rosoid, and *Cotoneaster*, a maloid.

Selected Genera of Rosaceae

Alchemilla • Amelanchier • Aruncus • Cliffortia • Cotoneaster • Crataegus • Cydonia • Eriobotrya • Filipendula • Geum • Gillenia • Holodiscus • Kerria • Lyonothamnus • Malus • Mespilus • Potentilla, including *Fragaria • Pyracantha • Pyrus • Rhodotypos • Rosa • Rubus • Sanguisorba • Sorbaria • Sorbus • Spiraea*

RUBIACEAE

Madder or Coffee Family

About 12,000 species in 575 genera

RANGE Cosmopolitan, but best developed in warm temperate and tropical climates

PLANT FORM Large and small trees, shrubs, rarely vines, also herbaceous perennials and some annuals; leaves opposite or in whorls, always simple, entire, sometimes small, occasionally needle-like, usually with stipules present between the leaf petioles and appressed to stem

FLOWERS Usually bisexual, radially symmetric or zygomorphic, usually in cymes (or solitary); calyx mostly of 4 (or 5) partly united sepals, green or petaloid; corolla mostly 4 (or 5, as many as sepals, rarely 0–8) petals united at least basally or in a tube with spreading limbs, the tube often well developed; stamens usually as many as petals, inserted on corolla tube and alternating with the petals, splitting longitudinally; ovary inferior, of 2–5 united carpels, each with 1 to many ovules, with terminal style (free styles in *Galium*), each carpel with 1 to many ovules

FRUIT Variously a dry capsule, drupe, or berry

A large and overwhelmingly tropical family, Rubiaceae are usually divided into four subfamilies: Antirheoideae (*Alberta, Fadogia*), Cinchonoideae (*Cinchona* and many more tropical trees), Dialypetalanthoideae (*Coffea, Gardenia, Pavetta, Pygmaeothamnus*, and many more), and Rubioideae (*Coprosma, Galium, Oldenlandia, Pentas, Rubia*). The family is more important in tropical than temperate areas for horticulture. *Gardenia* is widely grown for its strongly perfumed, large, white flowers. The red-flowered trees *Alberta* and *Burchellia* are grown as ornamentals in gardens with warm temperate or Mediterranean climates. Some species of the Asian genus *Leptodermis* are attractive, hardy shrubs. *Galium odoratum* (sweet woodruff) is a common ground cover although the white flowers are quite small.

Coffee, made from the roasted seeds of *Coffea arabica* and *C. robusta*, is perhaps the most significant product for humans, but Rubiaceae are also the source of the antimalarial drug quinine (from *Cinchona* bark) and the emetic ipecac (from *Carapichea* roots and rhizomes). Many trees provide timber in the tropics, none of particular importance. *Rubia tinctorum* is the source or the red dye madder, often substituted today by synthetics but still important in dyeing wool for oriental carpets.

Woody members of the family are immediately recognizable by their opposite, simple leaves with interpetiolar stipules appressed to the stem, four-merous flowers with a corolla tube, inferior ovary, and stamens alternating with the corolla lobes. Herbaceous members of the family are confusing as they may have leaves in whorls (cleavers or goose grass, *Galium aparine;* sweet woodruff, *G. odoratum*), and the stipules are leaf-like and hardly distinguishable from true leaves.

Selected Genera of Rubiaceae

Alberta • Anthospermum • Burchellia • Canthium • Carapichea • Cinchona • Coffea • Coprosma • Fadogia • Galium • Gardenia • Ixora • Morinda • Mussaenda • Nenax • Oldenlandia • Pavetta • Pentas • Psychotria • Rothmannia • Rubia

Rubiaceae, *Gardenia cornuta*. Natal gardenia is a shrub or small tree with simple, opposite leaves clustered on spur shoots that are characteristic of the family in their conspicuous interpetiolar stipules, forming a sheath around the stems. The large waxy flowers are solitary at the branch tips, with a long corolla tube and an inferior ovary that ripens into a woody or leathery berry crowned with the persistent remains of the calyx. Although Rubiaceae typically have four- (or five-) lobed flowers, those of some gardenias, like several other moth-pollinated flowers, are more dissected than usual, with a six- to eight-lobed perianth. The style terminates in a ribbed, club-shaped stigma.

Rubiaceae, *Pentanisia prunelloides*

Rubiaceae, *Pygmaeothamnus chamaeodendrum*

Rubiaceae, *Pavetta revoluta*

Rubiaceae, *Burchellia bubalina*

RUBIACEAE   245

Rutaceae. *Murraya paniculata.* The imparipinnate leaves of *Murraya* have alternate leaflets, and the leaves and fruits are dotted with the aromatic oil glands that characterize the family. The white, highly fragrant flowers have five small sepals, five free, waxy petals, and ten free stamens with slightly flattened filaments in two unequal whorls. Each anther is tipped with an oil gland. The ovary is situated on a fleshy, cushion-like nectary, with a solitary style and swollen stigma, and matures into a small berry.

RUTACEAE

Rue or Citrus Family

About 1975 species in 155 genera

RANGE Cosmopolitan, especially tropics, also many in southern Africa and Australia

PLANT FORM Trees and shrubs, a few perennial herbs, sometimes thorny, mostly evergreen and usually aromatic; leaves usually spirally arranged, often compound, pinnate or trifoliolate, or simple then sometimes deeply lobed or small to minute and ericoid, with pellucid gland dots, without stipules

FLOWERS Usually bisexual and radially symmetric (rarely zygomorphic), in cymes, racemes, or solitary, some unisexual with male and female flowers on different plants; calyx with (2 to) 5 free or basally united sepals, usually green; corolla usually with 4 or 5 free or partly united petals; stamens in 2 whorls, sometimes with 1 whorl sterile or absent, or 3–4 times as many as petals, sometimes filaments partly or fully united, anthers with longitudinal slits; ovary superior, of 2–4 (or 7) united carpels (or sometimes more or less free), thus 2- to 4- (or 7-) celled, with 1 terminal style or each carpel free distally and with separate terminal style, each carpel with several to many ovules

FRUIT Usually dry with carpels separating at maturity, or fleshy berries or drupes

Almost always aromatic, sometimes unpleasantly so, Rutaceae offer relatively few species for temperate gardens, but *Ruta graveolens* (rue) is commonly cultivated for its fine blue-gray foliage although its flowers are inconspicuous. Several species (or only one according to some authorities) of *Dictamnus* (dittany), constituting one of the few herbaceous genera in the family, are grown for their large, attractive, white or pink flowers, also unusual in Rutaceae in being zygomorphic. Also known as burning bush, the volatile oils of *D. albus* may accumulate around the plant on sultry summer days to such a concentration that they ignite if a flame is held nearby.

Several *Citrus* species and cultivars are half-hardy in U.S. Department of Agriculture Plant Hardiness Zone 7 but thrive in areas of Mediterranean or frost-free warm temperate climate, including oranges, lemons, mandarins, satsumas, and their ilk. The African *Calodendron capense* (Cape chestnut) is a particularly attractive tree with large pink flowers borne in profusion, and its smaller relatives from the Cape Region of South Africa offer several striking shrubs. Most notable of these are *Adenandra* (porcelain flower), *Coleonema* (confetti bush), and several species of *Agathosma* (buchu), all small to moderate-sized heath-like shrubs with aromatic foliage.

The eastern Asian genus *Skimmia*, which includes *S. japonica* and *S. laureola*, both hardy evergreen shrubs, are widely cultivated in temperate gardens for their red berries, borne in the winter months. Female plants bear the berries,

Rutaceae, *Adenandra villosa*

Rutaceae, *Chiosya ternata*

Rutaceae, *Dictamnus albus*

the male plants having clusters of rather insignificant white flowers. The Mexican and southwestern North American *Choisya arizonica* and *C. ternata* are perhaps the most hardy of the shrubby Rutaceae, and their hybrid, the cultivar 'Aztec Pearl', is especially fine. Like all Rutaceae, *Choisya* has aromatic foliage though perhaps not the most pleasant. Some cultivars have attractive golden yellow foliage useful in landscaping. Hardy only in Mediterranean and warmer climates, *Correa* and *Murraya* (for example, *M. paniculata*, orange jessamine) have several attractive species in cultivation.

The family is of major economic importance with some species and many cultivars of the genus *Citrus* grown in orchards: *C. sinensis*, the orange, is by far the most important. Lemons are the hybrid *C. aurantium* × *C. medica* (called *C. ×limon*), and limes are *C. ×aurantifolia*. The pomelo is yet another species, *C. maxima*, and the grapefruit is a hybrid of this species and *C. reticulata*. Mandarins, tangerines, and satsumas are cultivars of *C. reticulata*. The flavoring in Earl Grey tea is obtained from bergamot oil, another citrus cultivar. Breeding of citrus fruit continues today in the search for new strains for food and medicinal uses.

Other Rutaceae in cultivation are species of *Agathosma*, especially *A. betulina*, grown for its oils used for flavorings and its medicinal properties. Szechuan pepper, obtained from bark and berries of species of *Zanthoxylum* native to China, is used in cooking and medicinally. The rind of several *Citrus* species produces bitter agents (as in tonic water) and insecticides, some compounds being altered to form sweeteners.

Selected Genera of Rutaceae
Adenandra • Agathosma • Boronia • Calodendron • Choisya • Citrus • Coleonema • Correa • Dictamnus • Diosma • Fagara • Murraya • Ruta • Skimmia • Zanthoxylum

SALICACEAE

Willow or Poplar Family

About 1250 species in 54 genera

RANGE Cosmopolitan, notably northern temperate to Arctic, but many tropical

PLANT FORM Trees and shrubs, some trailing, many deciduous; leaves simple, serrated, in spirals, usually with stipules

FLOWERS Bisexual or unisexual (*Populus*, *Salix*), usually radially symmetric and small but occasionally zygomorphic, in terminal or axillary racemes or heads (spikes or catkins in *Salix* and *Populus*); calyx mostly of 3–8 sepals, usually green; corolla with as many free petals as sepals or lacking; stamens (1 or) 2 in most *Salix* species to many, sometimes connate in groups; ovary superior, compound, of 2–13 united carpels, mostly 1-celled, with 1–8 separate styles, each carpel with several to many ovules

FRUIT A berry, drupe, or capsule; in wind-dispersed *Populus* and *Salix*, seeds minute with tuft of woolly hairs (hence cottonwood)

The circumscription of Salicaceae has been expanded, largely as a result of DNA sequence studies, to include genera of the family Flacourtiaceae and two others of no horticultural significance. The expanded Salicaceae now comprise several more genera as well as *Populus* (aspens and poplars) and *Salix* (sallows, willows). Among these is *Azara*, a South American genus of evergreen trees of some ornamental value; *A. microphylla* from Argentina and Chile is the most hardy and has attractive, vanilla-scented yellow flowers produced in winter. A second addition to Salicaceae, the eastern Asian *Idesia*, is a deciduous tree bearing drooping clusters of red berries up to a third of an inch (about 8 mm) in diameter. Several species of *Salix* are grown in gardens and parks, perhaps most obviously the weeping willow, grown on the banks of streams. This is *S. babylonica* or its hybrid with *S. alba* called *S. ×sepulcralis*. The "willows" under which the children of Israel wept in their Babylonian exile are probably *Populus euphratica*.

Willow, specifically the bark of *Salix alba*, was the natural source of salicin, which is converted in the body to salicylic acid, the active ingredient of aspirin. This remains an important analgesic and blood thinner but is today synthesized industrially. Wood of various willows has numerous uses, especially basketry and cane furniture. *Populus* species are grown in plantations for wood pulp for manufacture or paper.

Selected Genera of Salicaceae
Azara • Flacourtia • Homalium • Idesia • Populus • Salix • Xylosma

Salicaceae, *Salix apoda*

SAXIFRAGACEAE

Saxifrage Family

About 640 species in 34 genera

RANGE Mainly northern temperate, Eurasia and North America, a few extending though the Andes to southern South America

PLANT FORM Mostly perennials, a few annuals, many more or less succulent, some with large underground rootstocks; leaves simple, then sometimes highly dissected (*Astilbe*), or palmately to pinnately compound, often in basal rosettes, or spirals, without stipules

FLOWERS Bisexual, usually radially symmetric but occasionally zygomorphic, in cymes, racemes, or solitary; calyx with (3–)5(–10) sepals, usually green; corolla with free or partly united petals, as many as sepals; stamens in 2 whorls or with 1 whorl sterile or absent; ovary superior, compound, of 2 (or 3) united carpels (rarely more or less free), thus 2- or 3-celled, each carpel free distally and with separate terminal style, each carpel with several to many ovules

FRUIT Usually dry capsules, splitting completely or through pores

Current understanding of the relationships of many erstwhile genera of Saxifragaceae indicates that their affinities lie elsewhere. For example, *Parnassia* has been referred to Celastraceae, and *Deinanthe* and several other genera are included in Hydrangeaceae. Gooseberries and currants (*Ribes*) are included in Grossulariaceae (see the appendix, Genera of Small Families Otherwise Not in General Cultivation). These changes render Saxifragaceae more homogeneous and easier to recognize. Most common in temperate gardens are species of saxifrage (*Saxifraga*), *Astilbe*, and *Bergenia*. The genus *Heuchera* is becoming more common in gardens as hybrids and cultivars with striking leaf colors and patterns have become available.

The family is distinguishable by its mostly two-celled ovary, the carpels sometimes free distally, each with a separate style, and often succulent leaves. Genera

Saxifragaceae, *Bergenia cordifolia*

such as *Astilbe* are easily confused with Rosaceae (for example, *Aruncus*, goat's beard, or *Filipendula*, burnet or queen of the prairie) in leaf and flower because of their finely divided leaves and tiny flowers, but those genera have free carpels and stipules at the base of the leaves (neither easy to see).

Selected Genera of Saxifragaceae

Astilbe • *Astilboides* • *Bergenia* • *Boykinia* • *Chrysosplenium* • *Darmera* • *Heuchera* • *Mitella* • *Peltoboykinia* • *Rogersia* • *Saxifraga* • *Tellima* • *Tiarella* • *Tolmiea*

Saxifragaceae, *Tellima grandiflora*

Scrophulariaceae, *Phygelius capensis.* As in most Scrophulariaceae, the flowers of *Phygelius* have a calyx of five partially united sepals and a corolla in which the five petals are joined in a tube. *Phygelius* has four stamens, two longer and two shorter, and a superior ovary with a single style.

SCROPHULARIACEAE

Figwort Family

About 3700 species in 154 genera, including those also treated as Phrymaceae and Plantaginaceae

RANGE Mostly tropical and warm temperate, especially southern Africa, some cold temperate

PLANT FORM Mostly shrubs and herbs, including many perennials, some aquatics, and trees; leaves simple, mostly opposite, without stipules, sometimes tiny and appressed to stem like those of *Cupressus* (cypress)

FLOWERS Usually bisexual, usually zygomorphic and 2-lipped, sometimes radially symmetric, usually in spikes or racemes, sometimes thyrses, solitary, or in heads; calyx mostly 4- or 5-lobed (or 2- or 3-lobed), sepals free or partly united, usually green; corolla with petals partly united, sometimes tubular, mostly 4- or 5-lobed (or as many as 8-lobed), sometimes spurred basally; stamens 5 or 4, then lower pair often reduced (or 1, 2, or 3), inserted on corolla tube; ovary superior, compound, of 2 (or 3) united carpels and 2- (or 3-) celled, with 1 terminal style, the stigma 2-lobed or capitate, each carpel with 1 to many ovules

FRUIT Usually dry capsules, splitting completely or through pores, rarely a berry or drupe

Change in the understanding of the composition and circumscription of flowering plant families has resulted in the separation of the families Phrymaceae and Plantaginaceae (including Veronicaceae) from Scrophulariaceae, in which their genera were historically included. The families differ in small details or hardly at all; in particular, Phrymaceae and Plantaginaceae often have the lobes of the calyx united for some distance, and flowers of some Plantaginaceae may have just two stamens (for example, *Veronica*). For convenience we treat all three families here but note below genera that fall into each of the three.

The genus *Paulownia*, sometimes included in Bignoniaceae, may also be treated as the separate family Paulowniaceae. Although a fast-growing tree with attractive flowers carried in large panicles, *P. tomentosa* is invasive in forests of eastern North America. The diagnostic characters closely match those of Bignoniaceae except that the leaves are invariably simple. *Paulownia* is best regarded as a member of Scrophulariaceae.

Scrophulariaceae and Plantaginaceae include many garden favorites, notably *Antirrhinum* (snapdragon), *Digitalis* (foxglove), *Linaria* (toadflax), *Verbascum* (mullein), and several species of *Veronica* (speedwell) and *Veronicastrum*. Several species of *Plantago* (plantain) are bothersome garden weeds. A few species of the perennial genus *Chelone* (turtle flower) are occasionally seen in gardens.

Plantaginaceae, *Veronica syriaca*

Scrophulariaceae, *Diascia vigilis*

Scrophulariaceae, *Verbascum saccatum*

Scrophulariaceae, *Nemesia affinis*

Phrymaceae, *Diplacus aurantiacus*

The relatively small family Phrymaceae include two genera of horticultural merit: *Mimulus* (monkey flower), which includes annuals and perennials, and *Diplacus*, which are shrubs. Several species are grown in temperate gardens, especially those in areas of Mediterranean climate. The sticky monkey flowers include *D. aurantiacus*, which is perhaps the best known species of the genus.

Members of Scrophulariaceae in the broad sense employed here are generally recognizable by their corolla with lobes united below and often two-lipped, calyx lobes likewise often partly united, and superior ovaries of two united carpels, these maturing into more or less heart-shaped or conical capsules. Stamens range in number from five or four to only two in the *Veronica* group of genera. The parasitic genera formerly included in Scrophulariaceae, among them *Castilleja* (Indian paintbrush), *Orobanche* (broomrape), and many more, are now assigned to the family Orobanchaceae. Few are ever deliberately grown in gardens.

Calceolaria (slipper flowers), sometimes included in Scrophulariaceae, is now considered sufficiently distinct to be in its own family, Calceolariaceae (described following Bignoniaceae).

The shrubby genera *Hebe*, of Australasian origin, especially New Zealand, and *Parahebe* have the diagnostic floral characters of *Veronica* and are now included in that genus, much to the confusion of gardeners and horticulturists, most of whom prefer to keep them apart. Species of *Scrophularia* are sometimes grown in gardens. Among the African genera of the family, *Diascia* (twin spur) and *Nemesia* (Cape snapdragon) are more widely grown as attractive selections of both annual and perennial species become available in the nursery trade. *Phygelius* (for example, *P. capensis*, so-called Cape fuchsia) is likewise receiving horticultural attention. A few species of the widespread genus *Buddleja* are hardy and are grown locally, but *B. davidii* (sometimes called butterfly plant) is a noxious and invasive weed in Europe and North America, although several cultivars are regrettably still grown in gardens.

Selected Genera of Phrymaceae
Diplacus • Mimulus

Selected Genera of Plantaginaceae
Angelonia • Antirrhinum • Bacopa • Chelone • Cymbalaria • Craterostigma • Digitalis • Ellisiophyllum • Linaria • Ourisia • Penstemon • Plantago • Rehmannia • Synthris • Torenia • Veronica, including *Hebe* and *Parahebe • Veronicastrum*

Selected Genera of Scrophulariaceae
Alonsoa • Buddleja • Diascia • Nemesia • Paulownia • Phygelius • Scrophularia • Sutera • Verbascum

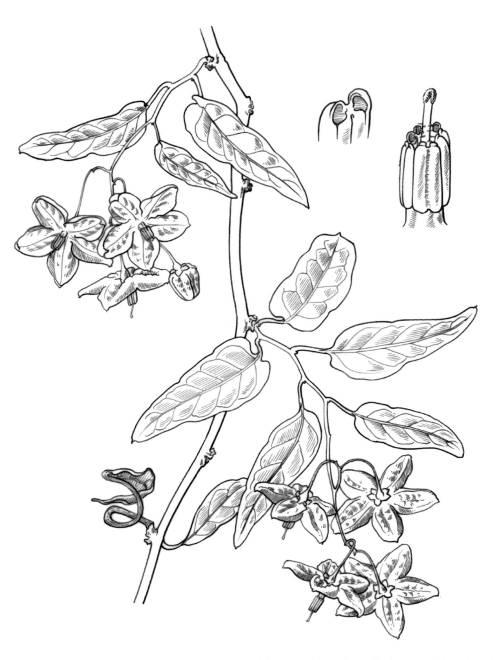

Solanaceae, *Solanum laxum*. The flowers of nightshade are typical of many members of the family, including potato and tomato, in having five sepals, five petals united into a short tube, and five stamens that converge but are not fused over the ovary. Anthers in *Solanum* dehisce by means of apical pores, but in other genera dehiscence is more conventional, by longitudinal slits.

SOLANACEAE

Nightshade or Potato Family

About 2650 species in 90 genera

RANGE Nearly cosmopolitan but especially warm temperate and tropical North and South America

PLANT FORM Some trees, many shrubs, vines, herbaceous perennials and annuals, all parts often hairy (with branched hairs); leaves simple, entire to lobed or compound, some trifoliolate, usually in spirals, without stipules

FLOWERS Usually bisexual and radially symmetric, some zygomorphic, in cymes or solitary; calyx of 5 sepals virtually free or variously united, persistent (enclosing fruit in *Physalis*), sometimes colored; corolla mostly of 5 (or 4 or 6) petals united basally or in a short to long tube; stamens usually 5, attached to corolla, sometimes only 4, or 2 with 2 sterile, anthers often converging but not fused, with longitudinal slits or apical pores; ovary superior, of 2 united carpels, 2- to 4-celled, usually with many ovules, style terminal, 1, with capitate or lobed stigma

FRUIT Dry, many-seeded capsules, or berries, these sometimes large as in the tomato and eggplant

A relatively large family, Solanaceae have contributed substantially to ornamental horticulture. Annuals seen in temperate gardens include *Petunia*, *Calibrochoa*, *Salpiglossis*, *Schizanthus*, and some species of *Nicotiana*, the genus of tobacco, many with unusual flowers. Less often encountered are the tropical *Browallia* and *Nierembergia*. The shrubby genera *Brunfelsia*, which includes *B. australis* (yesterday-today-and-tomorrow), and *Cestrum* are barely cold hardy or not at all, but many *Cestrum* species can be grown as annuals. *Brunfelsia* is perfectly at home in warm temperate and Mediterranean gardens. *Brugmansia* (angel's trumpets), with its huge, pendent, white, pink, or orange flowers, is not hardy but can be over-wintered in the greenhouse or, with care, maintained indoors.

More significant to humans are the food plants contributed by Solanaceae. The potato, *Solanum tuberosum*, is a major staple in much of the world, and tomato, *S. lycopersicum*, is almost equally important. Another *Solanum* species that contributes to the human diet is *S. melongena* (aubergine, brinjal, or eggplant). The genus *Physalis* contributes *P. philadelphica* (tomatillo) and *P. peruviana* (Cape gooseberry or ground cherry). Peppers and various chilies are the fruits of cultivars of *Capsicum annuum*. Tobacco is obtained from dried leaves of *Nicotiana tabacum*.

Many Solanaceae contain alkaloids and other poisonous compounds, nicotine among them. *Datura*, sometimes called jimson weed (mainly applied to *D. stramonium*), is highly toxic but also used medicinally and as a hallucinogen; *D.*

Solanaceae, *Hyoscyamus aureus*

Solanaceae, *Solanum crispum*

Solanaceae, *Solanum panduriforme*

Solanaceae, *Lycium afrum*

stramonium is invasive in some parts of the world. The genus *Atropa* is the source of atropine, and *A. belladonna* (deadly nightshade) was used in the past by women to dilate their pupils, making their eyes seem brighter. All parts of the plant are poisonous to humans. The Eurasian genus *Mandragora* (mandrake) is likewise poisonous, the roots fancifully resembling human form and said to shriek when uprooted and deafen gatherers. The plants contain hyoscyamine, a hallucinogen and soporific, used in the past as an anesthetic. The related genus *Hyoscyamus* is similarly toxic and contains similar compounds. *Hyoscyamus niger* (henbane) was used historically as a hypnotic, hallucinogen, and for pain relief, and more prosaically, to stupefy hens while being stolen.

Solanaceae are most easily recognized by their five-merous flowers with the corolla united for part of its length, sometimes tubular, the corolla lobes folded along the midline in bud, and a superior ovary of two united carpels and a single terminal style with capitate or lobed stigma. In many genera, including *Solanum* (for example, *S. laxum*, jasmine nightshade), the anthers are coherent and surround the style, the anthers then with terminal pores or slits. In these flowers the pollen is released only by vibrations caused by bee pollinators. This system, called buzz pollination, is particularly well developed in Solanaceae and occurs occasionally in many other plant families.

Selected Genera of Solanaceae
Atropa • Browallia • Brugmansia • Brunfelsia • Cestrum • Datura • Fabiana • Hyoscyamus • Iochroma • Lycium • Mandragora • Nicotiana • Nierembergia • Petunia • Physalis • Salpiglossis • Schizanthus • Solanum • Withania

STYRACACEAE

Styrax Family

About 170 species in 11 genera

RANGE Mostly warm temperate and tropical America, also the Mediterranean and Asia

PLANT FORM Mostly trees, some shrubs, often with star- or umbrella-like hairs (with the stalk toward the center); leaves simple, often with serrated margins, in spirals, without stipules

FLOWERS Usually bisexual, usually radially symmetric, in racemes or cymes without bracts; calyx of 4 or 5 (to 9) lobes or lacking, green; corolla with 4 or 5 (to 9) petals united basally; stamens 5 or twice as many as corolla lobes, often united in a tube around slender style, inserted on corolla tube; ovary superior or inferior, of 2–5 united carpels with 1 to many ovules, style 1, terminal

FRUIT Usually dry capsules or one-seeded drupes

Styracaceae are a small family but have several ornamentals, some only moderately known yet worth attention. *Styrax* includes the Mediterranean and Californian *S. officinalis*, well adapted to warm temperate gardens, while *S. japonica* and *S. obasia* are cold hardy and offer particularly attractive garden subjects in cool and warm temperate regions. *Halesia* includes the spring-blooming *H. carolina* and *H. monticola*, the Carolina silverbell trees. Less well known are the trees *Pterostyrax* and *Rehderodendron*, both very ornamental, as is *Melliodendron*, which has large pink blooms produced before the leaves unfold.

Many *Styrax* species produce resins that yield benzoin, benzoic acid, and other volatile compounds of medicinal importance, including friar's balsam, and are used in many parts of the world in traditional medicine. Styracaceae are recognizable by their simple leaves, corolla of spreading petals united only near the base, stamens united around the style, and usually a superior (less commonly inferior) ovary.

Selected Genera of Styracaceae
Bruinsmia • Halesia • Huodendron • Melliodendron • Pterostyrax • Rehderodendron • Sinojackia • Styrax

EBENACEAE The ebony family, related to Styracaceae and Ericaceae, comprise two genera and some 575 species, nearly cosmopolitan in distribution but few in temperate or cold climates. Consistent with these other families, the leaves of Ebenaceae are simple and spirally arranged, the sepals and petals partially united, the petals curved back distally, and the stamens are inserted on the corolla tube. The calyx is persistent and often accrescent, thus enlarged in fruit, a feature evident in the large, fleshy, berry-like fruit. The genus *Diospyros* includes *D. virginiana* (common persimmon), a North American native that has edible fruit, ripening and turning bright orange after the onset of cold weather. The eastern Asian *D. kaki* (Japanese or Chinese persimmon) has particularly large orange fruits in which the accrescent calyx is very evident. Often seen in markets in early winter, these persimmons are much favored in China, Japan, and adjoining countries. Many cultivars of both species produce fruit without seeds. Extremely hard, ebony wood is obtained from several large trees of the genus, with *D. ebenum* from India and Sri Lanka the original ebony of commerce.

Styracaceae, *Styrax japonica*

Styracaceae, *Styrax officinalis*

Theaceae, *Camellia* cultivar. Camellias are evergreen, have leathery leaves, and like other members of the family have free petals and numerous stamens. The superior ovary has a single style that is branched at the tip.

THEACEAE

Tea or Camellia Family

About 225 species in 7 genera

RANGE Mainly tropical with a few warm temperate

PLANT FORM Trees and shrubs, mostly evergreen; leaves in spirals, simple, entire or toothed, without stipules

FLOWERS Perfect, usually large, radially symmetric, solitary in axils or terminal; calyx with 5 (or 6 or 7) sepals sometimes united basally, green; corolla mostly of 5 petals (many petals in *Camellia* cultivars), free or basally united, thus the corolla falling as a unit; stamens usually many (or 5 or 10), free or basally united, sometimes in 5 clusters, each attached to petals, anthers with longitudinal slits; ovary superior, of 3–5 (to 10) united carpels with as many cells, styles separate or united basally

FRUIT A capsule with persistent central column, or a drupe

Theaceae, *Camellia sasanqua*

Theaceae, *Stewartia serrata*

Most temperate gardens contain at least one member of Theaceae, the genus *Camellia*. Camellias have been cultivated in China and Japan for centuries, and there are numerous cultivars, most with doubled flowers with many petals and sometimes lacking stamens. Wild species have rather smaller flowers with just five petals and multiple stamens. *Camellia thea* (in the past placed in a separate genus, *Thea*) is the source of the beverage tea. Tea plants are cultivated in plantations across the world where climate is suitable. A second genus of horticultural importance, *Stewartia*, includes several species of deciduous trees and shrubs in eastern Asia and two in eastern North America, a disjunction repeated in many plant families. *Stewartia* species have large, ephemeral flowers, and many have attractive bark and usually striking autumn foliage in shades of red and orange.

Also native to eastern North America, the genus *Franklinia* has just the one species, *F. alatamaha*, discovered in the 18th century in Georgia and named in honor of Benjamin Franklin. Cuttings planted in Philadelphia soon after the discovery of the species

are the source of all the plants in cultivation. Repeated searches have failed to find any sign of the small tree, now almost certainly extinct in the wild. The tree has the typical attributes of Theaceae, large flowers and numerous stamens, and like *Stewartia* is deciduous and offers brilliant autumn color.

Tea is the only significant product of Theaceae for humans. Young leaves are harvested by hand and prepared in various ways to produce white, green, or black tea; for the latter the leaves are allowed to wilt and then fermented before drying. Tea contains several alkaloids, these responsible for its stimulating effect, including caffeine, theobromine, and theophylline. Oil obtained from seeds of *Camellia oleifera* and *C. sasanqua* are used in China and Japan for cooking, cosmetics, and lubricants. The family is otherwise important as ornamentals and a minor source of timber.

Genera of Theaceae
Apterosperma • *Camellia,* including *Thea* • *Franklinia* • *Gordonia* • *Schima* • *Stewartia* • *Ternstroemia*

Thymelaeaceae, *Lasiosiphon rigidus.* The floral tube in Thymelaeaceae, interpreted as a hypanthium, bears four or five petal-like sepals; true petals are either lacking or reduced to minute scales. The stamens are inserted on the hypanthium, often in two whorls of four or five each. The style in *Lasiosiphon* is characteristic of several genera in its eccentric insertion to one side of the top of the ovary.

THYMELAEACEAE

Daphne Family

About 850 species in 45 genera

RANGE Cosmopolitan, mostly tropical and subtropical and areas of Mediterranean climate

PLANT FORM Large and small trees and shrubs, rarely herbs or vines, with tough, fibrous bark tearing in strips; leaves simple, sometimes needle-like, entire, mostly in spirals, rarely opposite, without stipules; often spring or summer blooming, even in winter (*Daphne bholua*, *D. laureola*)

FLOWERS Usually bisexual, small and radially symmetric, borne at branch tips or on short shoots in racemes or small to large head-like clusters; perianth seemingly consisting of 1 whorl but interpreted as a hypanthium with petaloid sepals on the tube rim, 4- or 5-lobed, variously colored but often pink or mauve, sometimes even green; corolla reduced to scales as many as sepals or absent; stamens as many as or twice as many as sepals, inserted within or at rim of hypanthium; ovary superior, compound, most often of 2 or more closely united carpels, 2- or 1-celled with 1 ovule per cell, style 1 and often eccentric

FRUIT Often fleshy, berry-like (especially *Daphne*), sometimes dry capsules

The family includes *Daphne*, several species of which are grown in northern temperate gardens, partly for their deliciously scented flowers, especially *D. bholua* and *D. odora*. The Mediterranean *D. mezereum* has flowers borne on woody stems. Green-flowered *D. laureola* is invasive in parts of northwestern North America. The deciduous *Edgeworthia* (Chinese paper tree) is an early spring bloomer with fine, rounded heads of yellow or orange flowers. Several species of the shrubby African genus *Gnidia* (and the related *Lasiosiphon*, for example, *L. rigidus*, saffron bush) and the Australian *Pimelea* are favorites in southern-hemisphere gardens and are grown elsewhere but are not cold hardy. The small southern African tree *Dais cotinifolia* is widely grown in warm temperate climes and is used in street plantings.

Thymelaeaceae, *Daphne sericea*

Thymelaeaceae, *Dais cotinifolia*

The fleshy fruits of *Daphne* are poisonous to humans and many mammals but relished by birds, which distribute the seeds. Several species have medicinal uses. The Chinese *D. genkwa* contains an effective and safe abortifacient. The bark of some *Daphne* species is used to make rope and paper, and that of *Edgeworthia* is used in China to make a high-quality paper. The East African *Gnidia glauca* is another source of fine-quality paper.

Thymelaeaceae are recognizable by their simple leaves and small, radially symmetric, four- or five-merous flowers with the perianth evidently of one whorl, actually a hypanthium with calyx lobes on the rim and petals if present reduced to small scales. Fruits are often fleshy and poisonous (especially *Daphne*). The exceptionally tough bark is hard to break and tears in strips when branches are broken, a useful feature for identification.

Selected Genera of Thymelaeaceae
Dais • Daphne • Edgeworthia • Gnidia • Lachnaea • Ovidia • Passerina • Pimelea • Struthiola • Thymelaea

VERBENACEAE

Verbena Family

About 1000 species in 32 genera

RANGE Tropics, especially South America, a few subtropical or warm temperate

PLANT FORM Trees, shrubs, vines, and some herbs, young stems often four-sided; leaves usually simple, opposite or whorled, serrated to finely dissected, without stipules, sometimes aromatic when crushed (especially *Lippia*)

FLOWERS Perfect, zygomorphic (mostly 2-lipped) or very nearly radially symmetric, in racemes, cymes, or crowded in heads; calyx 4- or 5-lobed with sepals partly united, sometimes enlarged, green or colored; corolla 4- or 5-lobed, with petals partly united in a narrow or wide tube; stamens 4 (sometimes 1 more a staminode), inserted on corolla tube, anthers with longitudinal slits; ovary superior, compound, of 2 closely united carpels, initially 2-celled but divided by intrusive partitions into 4 cells, each with 1 ovule, style 1

FRUIT Consisting of four single-seeded, nutlet-like sections, separating when ripe, or fleshy drupes

The circumscription of Verbenaceae has changed as a result of molecular studies and continues to do so. As a consequence, many genera once included in the family have been removed, mostly to the closely related Lamiaceae (sage or mint family), notably *Callicarpa* (beauty bush), *Clerodendrum*, *Teucrium* (germander), and *Vitex*. The genus *Tectona*, which includes the tropical Asian *T. grandis* (teak) has likewise been removed to Lamiaceae. Thus readers of older works about plant families will find little in common with this current account of Verbenaceae.

Verbenaceae are distinguishable from most Lamiaceae by their terminal style, but the distinction is unreliable as a fair number of genera of Lamiaceae also have a terminal style. Most Verbenaceae have quite small flowers aggregated into spikes or head-like clusters, not frequent in Lamiaceae. Verbenaceae familiar to gardeners include *Verbena*, especially *V. bonariensis*, which is also weedy in parts of the world, and *Glandularia*, the verbena of horticulture. *Lippia* species are sometimes grown in gardens, one species as an unusual lawn. *Lippia dulcis* contains hernandulcin, a compound many times sweeter than sucrose. The vine *Petrea volubilis* is grown in warm temperate and tropical gardens, the attractive bright to dark blue calyx lobes that surround each flower persisting long after the blue corolla has faded and fallen. *Lantana* comprises several shrubby species not truly hardy but often grown in warm temperate and Mediterranean gardens for their attractive heads of tiny flowers. *Lantana camara* is an aggressive weed in parts of the world, the fleshy fruits spread by birds. *Lantana* species are sometimes grown as annuals in temperate gardens or overwintered in greenhouses.

Verbenaceae are recognizable by their simple, opposite leaves without stipules, four- or five-merous flowers with sepals and petals each partially united, four stamens (or fewer), and a superior, two-celled ovary later divided into four cells, each with a single ovule. The style is always single and terminal. This is more or less the same as Lamiaceae, which often but not always have a basal style. The pagoda-like arrangement of several flowers at nodes of the flowering stems of many Lamiaceae is rarely found in Verbenaceae, which usually have flowers crowded in terminal clusters. The virtually identical circumscriptions of the two families sometimes makes it necessary simply to learn which genera belong in which family.

Selected Genera of Verbenaceae

Aloysia • *Chascanum* • *Duranta* • *Glandularia* • *Lantana* • *Lippia* • *Petrea* • *Priva* • *Stachytarpheta* • *Verbena*

VIOLACEAE

Violet Family

About 815 species in 29 genera

RANGE Cosmopolitan, only *Viola* temperate

PLANT FORM Trees, shrubs, and vines but mostly perennial herbs; leaves simple, entire to dissected, usually in spirals, with stipules

FLOWERS Perfect, radially symmetric or zygomorphic, in racemes, panicles, or solitary in leaf axils; calyx of 5 free sepals, sometimes persistent; corolla of 5 free petals, when zygomorphic lowermost petal spurred; stamens (3 or) 5, free or filaments united, anthers sometimes wrapped around ovary; ovary inferior, of (2 or) 3–5 united carpels, 1-celled, with 1 style

FRUIT Dry capsules or fleshy and berry-like; seeds sometimes with fleshy appendages (elaiosomes)

Violaceae, *Viola pygmaea* (from Carmen Ulloa Ulloa)

Violaceae, *Viola pygmaea* (from Carmen Ulloa Ulloa)

A family of moderate size, Violaceae include just one genus important in horticulture, *Viola* (pansies and violets), several species of which are grown in temperate gardens. Comprising nearly half the species in the family, *Viola* has flowers with the lower petal forming a landing for insects and a spur containing nectar. *Viola tricolor* (heart's ease, Johnny-jump-up) is widely cultivated and is a parent of cultivated pansies, which have a complex history of hybridization with other species. The scented violet, *V. odorata*, is the source of an essential oil used in flavorings and perfume production, and the crystallized flowers are used as food decorations.

Selected Genera of Violaceae
Alexis • *Corynostylis* • *Hybanthus* • *Hybanthopsis* • *Melicytus* • *Rinorea* • *Viola*

Two families related to Violaceae, **Hypericaceae** (St.-John's-wort family) and **Linaceae** (flax family), each include several genera but just one each of which is significant in temperate horticulture. In Hypericaceae, several species of the genus *Hypericum* (the name

St.-John's-wort is loosely applied to any species of the genus) are cultivated, all with yellow, four- or five-merous flowers, numerous stamens, simple, opposite leaves, and a superior ovary, features fairly typical of other genera of the family.

Linum is the only genus of Linaceae found in gardens; the family and genus have five-merous flowers with free petals and a superior ovary with styles separate or deeply divided. *Linum* includes several ornamental species with yellow, blue or red flowers. Cultivars of *L. usitatissimum*, a cultigen domesticated more than 7000 years ago in the Middle East, is the source of flax for fiber, especially linen (the word is obviously linked with the scientific name), used in textiles and rope. The seeds are edible and the source of linseed oil. Other species of *Linum* have medicinal uses and also have edible seeds.

VITACEAE

Grape Family

About 900 species in 16 genera

RANGE Mainly tropics and subtropics, a few warm temperate, often in arid habitats

PLANT FORM Evergreen or deciduous vines and tree-like plants with large trunks and succulent leaves; leaves digitately compound or simple, then often with palmate lobes or venation, sometimes succulent, arranged in spirals, sometimes with tendrils opposite the leaves, with stipules, these often inconspicuous or soon deciduous

FLOWERS Bisexual or not, radially symmetric, in cymes or panicles, often terminal or opposite leaves, mostly 4- or 5-merous; calyx small, often reduced to lobes or collar, green; corolla with small, usually green petals, sometimes united at tips and falling together; stamens 4 or 5, opposite petals; ovary superior, compound, mostly of 2 (rarely as many as 6) united carpels, 2-celled, each with 2 ovules, style 1 with capitate stigma, sometimes sunk in nectar-bearing disk

FRUIT A berry with two to four seeds

Vitaceae are best known to gardeners in northern climes as vines, including the grape, the genus *Vitis*, but the family also includes Boston ivy (*Parthenocissus tricuspidata*), native to eastern Asia, and Virginia creeper (*P. quinquefolia*), native to eastern North America. The vining members of the family have tendrils borne opposite the leaves, and in Boston ivy these terminate in small sticky disks that aid in climbing vertical rocks or buildings. Many cold-hardy Vitaceae have brilliant autumn color, sometimes but not always so in the domestic grape (*V. vinifera*).

Occasionally grown in gardens, the porcelain vine (*Ampelopsis brevipedunculata*), native to northeastern Asia, has foliage much like true grapes but attractively variegated green, pink, and white, both leaves and tendrils. The fruits turn dark blue when ripe but are not palatable as they contain large hard seeds with hardly any flesh. Nevertheless, they are attractive to birds, which disperse the seeds. The species has become an invasive weed in parts of North America. The southern African *Rhoicissus capensis* is occasionally grown in gardens there. Also southern African, *R. rhomboidea* is grown as an indoor plant in hanging baskets.

Other Vitaceae are plants of hot, dry habitats, sometimes forming small trees, often with disproportionately large, swollen trunks. Some are grown as houseplants or for display in greenhouses.

The grape, *Vitis vinifera*, is one of the oldest domesticated plants, originally providing food and wine, and later, brandies, sherries, and many more alcoholic beverages. The grape was probably native to the Middle East, whence it spread across the world wherever the climate was suitable. Dried grapes are called raisins (red), sultanas (white), or currants, the latter obtained from the Corinth grape, originally grown there (the word "currant" being a corruption of Corinth). True currants, black, red, or white, are species of the genus *Ribes* in the Grossulariaceae, not related at all to the grape family (see the appendix, Genera of Small Families Otherwise Not in General Cultivation).

The native North American *Vitis labrusca* is also grown for its fruit and provides an inferior wine, with a distinctive foxy or earthy flavor. Cultivars of *V. labrusca* include the Catawba and Concord grape. The native North American sap-sucking insect pest of American grape species, commonly called phylloxera, was accidentally introduced to Europe in the 19th century and in the 1860s began to devastate vineyards, as *V. vinifera* lacked resistance to the introduced insect. As a result, the French wine industry was all but destroyed, and vineyards elsewhere were also affected. *Vitis vinifera* grapes are now grafted onto native North American *Vitis* rootstocks that are largely resistant to phylloxera.

Vitaceae are easy to identify for gardeners in cool and warm temperate zones as the plants are vines with simple, palmately lobed and veined leaves, or digitately compound leaves; tendrils when present are borne opposite the leaves. The grape-like but small fleshy fruits are likewise distinctive. The flowers are so small that details are barely visible without magnification.

Selected Genera of Vitaceae
Ampelocissus • Ampelopsis • Cayratia • Cissus • Cyphostemma • Parthenocissus • Rhoicissus • Tetrastigma • Vitis

ZINGIBERACEAE

Ginger Family

About 1400 species in 50 genera

RANGE Mainly tropics and subtropics, a few warm temperate

PLANT FORM Evergreen or deciduous perennial herbs with creeping rhizomes, often aromatic; leaves simple, arranged in two ranks or spirals, with sheathing base and expanded blade, with parallel or pinnate veins and often prominent midrib, sometimes with an appendage (ligule) at the junction of blade and sheath, sheaths often wrapped around one another to form a pseudostem, without stipules

FLOWERS Perfect, zygomorphic, solitary or in spikes or heads, sometimes with prominent colored bracts; calyx of 3 green sepals, sometimes basally united; corolla tubular, with 3 unequal lobes spreading distally; stamens only 1 fertile, inner stamens forming a petaloid lip and upper 2 often also petaloid, filament of fertile stamen grooved on inner surface clasping style, anther with longitudinal slits; ovary inferior, compound, of 3 united carpels, 3-celled with many ovules (occasionally 1-celled), style 1, held in channel of fertile stamen, capitate or 2-lobed

FRUIT A dry capsule or fleshy and berry-like, sometimes with large hard seeds aromatic when pulverized

Few species of Zingiberaceae can be grown in temperate gardens, and those that are hardy are not often available in the nursery trade. At least some species of *Hedychium* (butterfly or ginger lilies) are seasonal perennials that tolerate freezing temperatures and merit horticultural attention for their large yellow, orange, or white flowers. Some *Hedychium* species are invasive in warm parts of southern Africa, New Zealand, and elsewhere. *Cautleya* and *Roscoea*, native to the Himalaya and western China, are well worth growing for their purple, mauve, or yellow flowers with an enlarged, petal-like lower lip, actually derived from the lowermost stamen. In warm temperate conditions, several other members of the ginger family are grown, including *Zingiber* (true gingers).

The rhizomes of several species of *Zingiber* are aromatic, and *Z. officinale* is widely grown as the source of ginger, used as a flavoring, either fresh or dried and pulverized, in savory or sweet foods. Other Zingiberaceae also yield spices essential in some cuisines. The dried and pulverized rhizomes of *Curcuma longa* provide the yellow spice turmeric, and other *Curcuma* species yield dyes. The spice cardamom is obtained from the seeds of the tropical Asian *Elettaria cardamomum*, and similar spices are obtained from seeds of *Amomum*. Seeds of several other genera are used locally as condiments, spices, and preservatives. The medicinal value of turmeric is uncertain but is used to treat various ailments.

Strelitziaceae, *Strelitzia reginae*. Like the related Zingiberaceae, Strelitziaceae have an inferior ovary, but the remarkable flowers, enclosed in a leathery, boat-shaped bract, have three large outer tepals and smaller inner tepals, one of which is enlarged and arrow-shaped with a channel on the inner surface.

Zingiberaceae, *Roscoea cautleyoides*

Zingiberaceae, *Roscoea*

The family is recognizable by its flowers, which although diverse in form have three green sepals, a tubular corolla with three unequal lobes, and petaloid sterile stamens often more prominent than the corolla lobes (thus evidently with as many as six petals). Flowers have a single fertile stamen, which is grooved on the inner (adaxial) side and clasps the style. Plants have creeping rhizomes, often thick, fleshy roots, and monocot-type leaves although often with pinnate venation, divided into a sheathing lower portion and expanded upper blade, the sheaths often clasping one another to form a false stem.

Selected Genera of Zingiberaceae

Aframomum • *Alpinia* • *Amomum* • *Cautleya* • *Costus* • *Curcuma* • *Elettaria* • *Hedychium* • *Kaempferia* • *Roscoea* • *Zingiber*

STRELITZIACEAE Families immediately related to Zingiberaceae include the largely tropical Musaceae (described under *Musa*, the banana, in the appendix, Genera of Small Families Otherwise Not in General Cultivation) and Strelitziaceae (bird of paradise family), which have similar leaves and growth habit but differ strikingly in floral details. Like Zingiberaceae, Strelitziaceae have an inferior ovary, but the outrageous flowers, enclosed in a leathery, boat-shaped bract, have three large outer tepals and smaller inner tepals, one of which is enlarged and arrow-shaped with a channel on the inner surface. The five or six stamens and the style are enclosed in the channel, more or less hidden from view. The South African *Strelitzia reginae* (bird of paradise or crane flower) is hardy in subtropical and Mediterranean climates, as is the elegant, tree-like Madagascan *Ravenala* (traveler's palm).

GENERA OF SMALL FAMILIES OTHERWISE NOT IN GENERAL CULTIVATION

ACORUS (Acoraceae) The genus *Acorus* and family Acoraceae have just two species of wetlands in North America and Eurasia. The flat leaves recall those of *Iris* (several species are called flags), hence the common name sweet flag. Several cultivars are grown, especially around ponds and streams, for their evergreen and often variegated foliage. The small, inconspicuous flowers are borne on a cylindric spike resembling the spadix of Araceae, to which the family is related. *Acorus* has floral parts in sets of three, but the flowers are so small that they are difficult to examine. The fruits are fleshy and resemble tiny berries. *Acorus* is the most archaic living genus of the monocots.

ASIMINA (Annonaceae) The eastern North American papaw (or pawpaw), also known as the custard apple, *Asimina triloba* is a small tree with large nutritious fruit and one of very few hardy members of the large, mostly tropical family Annonaceae. Historically, the papaw was an important food source for American Indians and was cultivated locally. Allied to Magnoliaceae, Annonaceae have simple leaves, a three-merous perianth, usually of three whorls of three tepals, the outer sometimes green and calyx-like, numerous stamens and carpels in spirals, and the carpels usually free (as in Magnoliaceae) but sometimes united. Several tropical species of the genus *Annona* are cultivated locally in the tropics for their edible fruit, including the cherimoya and custard apple (a name applied to the fruit of several Annonaceae).

CANNA (Cannaceae) Widely cultivated for their handsome foliage, dark red or variegated in several cultivars, *Canna* species are only half-hardy, although they may survive moderate freezing weather underground. Cannas thrive, of course, in areas of Mediterranean climate and in the tropics. The only genus of the tropical and warm temperate family Cannaceae, with about ten species of North, Central, and South America, *Canna* is recognizable by its large simple leaves with

long sheaths clasping the stem, red to yellow flowers with an inferior ovary, a single petaloid stamen with a pollen sac on one edge, and a petaloid style. The large, flaccid, petal-like organs are actually enlarged sterile stamens, which often conceal the smaller petals and calyx. As in most monocots, the ovary consists of three united carpels with three cells. Cultivars of *C. indica* or its hybrids with other *Canna* species are those most often seen in gardens. The specific epithet *indica* records its origin in the West Indies, not India.

COROKIA (Argyrophyllaceae) Belonging to the largely New Zealand but also Australian family Argyrophyllaceae, hardly known by name to northern-hemisphere gardeners, *Corokia* includes a handful of species suitable for warm temperate gardens. *Corokia cotoneaster* is perhaps best known to gardeners and is cultivated for its striking, some say bizarre, growth form. A shrub or small tree with tiny rhombic leaves, it has a zigzag branching pattern, resulting in an intricate and interesting open and diffuse appearance. The tiny flowers are bright yellow and when in bloom, *C. cotoneaster* is quite lovely. Like other families in the order Campanulales, the flowers have petals united in a tube. *Corokia* was formerly included in the dogwood family, Cornaceae, to which it is only distantly related.

CRINODENDRON (Elaeocarpaceae) A tropical and warm temperate family of some 600 species, Elaeocarpaceae include the southern South American genus *Crinodendron*. *Crinodendron hookerianum* (Chilean lantern tree) is a striking ornamental tree but although not cold hardy it survives in areas of Mediterranean climate. The buds of the drooping red flowers form the year before they open. Also Chilean, *C. patagua* (lily of the valley tree), has attractive, large, drooping white flowers with fringed petals. It tolerates some frost and is thus hardy in parts of Europe and North America. The family is recognizable by its simple, sometimes stipulate leaves arranged in spirals, pendent flowers with fringed petals, five to many stamens, and a superior ovary of as many as ten united carpels with a single style.

DROSERA (Droseraceae) Sundews have simple leaves bearing sticky glands that trap insects. The flowers are five-merous with free sepals and petals, a superior ovary, and prominent, usually bifid styles. *Dionaea* (Venus's-flytrap, also Droseraceae), a southeastern North American native occasionally grown indoors as a pot plant, has leaves with two hinged lobes bearing numerous interlocking teeth. When touched by an insect, the lobes close, trapping the visitor, which is slowly digested by secretions from surface glands.

EICHHORNIA (Pontederiaceae) The water hyacinth, sometimes grown in aquariums and ponds, is a floating aquatic monocot with attractive blue flowers; it has become a noxious weed of waterways in frost-free areas across the world. The flowers have two whorls of three tepals each, six stamens (some sterile), and a superior, three-celled ovary and a single style. Two more genera of the family, *Heteranthera* and *Pontederia*, are occasionally grown in pools.

ELAEAGNUS (Elaeagnaceae, oleaster family) Occasionally cultivated, *Elaeagnus* species are evergreen shrubs and trees with simple, spirally arranged leaves, distinctive in having silvery or brown scales over the surface. The small, inconspicuous flowers have a more or less inferior ovary.

ESCALLONIA (Escalloniaceae) *Escallonia rubra* is an evergreen shrub and hedge plant with deeply toothed leaves and small pink flowers. It is the only member of this poorly understood family in cultivation. Escalloniaceae are recognizable by simple leaves arranged in spirals, five stamens, and a superior ovary of two to four united carpels.

EUCRYPHIA (Cunoniaceae) Formerly the only representative of Eucryphiaceae, *Eucryphia* is now understood to be a member of the largely southern-hemisphere Cunoniaceae, a family of trees and shrubs. Leaves are opposite and usually compound and stipulate but simple in some of the several species of *Eucryphia*. The plants are grown in gardens for their especially beautiful, large, white flowers with four petals and multiple stamens, features exceptional in Cunoniaceae. Only half-hardy, *Eucryphia* species are widely grown in Australia and New Zealand, and in areas of mild, coastal, or Mediterranean climate in the northern hemisphere. Five species are native to Australia, and two to southern South America. *Cunonia capensis* is an attractive small tree occasionally grown in southern African gardens. The family has a superior ovary of two (to five) carpels with separate styles.

GUNNERA (Gunneraceae) The only genus of a family with some 40 species, *Gunnera* is typically found in wet situations. Several species are hardy, notably ground covers such as *G. magellanica;* others are very large stemless herbs. The ornamental *G. manicata,* perhaps native to Brazil or at least tropical South America, has huge, palmately veined leaves with spiny petioles, and impressive inflorescences of small flowers. It is sometimes grown in gardens with a mild temperate or Mediterranean climate. The Chilean *G. tinctoria* also has large leaves and is invasive in the British Isles, New Zealand, and elsewhere. The flowers are borne in large panicles with small female flowers below, even smaller male flowers above, and perfect flowers in the middle.

IMPATIENS (Balsaminaceae) Worldwide in distribution, the genus *Impatiens* includes some 1000 species, several used in temperate horticulture. Called impatience because of the explosive capsules that readily split when touched, flinging the seeds some distance, *Impatiens* includes both annuals and perennials, the latter best treated as annuals in cold climates as they do not tolerate frost. Several species are in cultivation, one of the most common being *I. wallichiana,* hundreds of thousands of which are sold annually for garden display. The flowers characteristically have a spur (derived from a sepal) and are often resupinate. The ovary is superior and enclosed by the fused anthers, forming a cap covering the stigma, a very unusual adaptation.

LARDIZABALA (Lardizabalaceae) The family, also with the genera *Akebia* and *Holboellia,* includes a few

Balsaminaceae, *Impatiens glandulifera*

hardy vines with palmately compound leaves and small, three-merous, unisexual or bisexual flowers with a superior ovary of 3 or 6–12 free carpels with sessile stigmas. The free carpels of this family signal a relationship with Ranunculaceae and related families. Lardizabalaceae have an unusual distribution—although centered in eastern Asia, *Lardizabala* is native to Argentina and Chile.

MELIA (Meliaceae, mahogany family) Mainly a tropical family of large trees, Meliaceae include *Azadirachta indica* (the neem tree) and *Melia azedarach* (Persian or Cape lilac, chinaberry), hardy at least in areas of Mediterranean climate. The family typically has compound, pinnate or bipinnate leaves. *Azadirachta indica* is the source of insecticidal neem oil, which also has medicinal properties, and its hardwood is a mahogany substitute. True mahogany is obtained from species of the genus *Swietenia* of the same family. *Melia azedarach* has attractive, scented flowers borne in large clusters, but it is invasive, the fleshy fruits being dispersed by birds into wild habitats where it

competes successfully with the native flora. Meliaceae typically have the compound leaves arranged in spirals, four- or five-merous flowers with filaments united into a tube, and a superior ovary.

MUSA (Musaceae, banana family) Tree-like, giant herbs with huge, entire leaves and fleshy, banana-like fruits, a few cultivars of *Musa* (banana) are hardy in Mediterranean climates but barely so in the temperate zone. The genus is mostly tropical or subtropical.

OXALIS (Oxalidaceae) A few species of the large genus *Oxalis*, mostly of Andean South America and South Africa, are cultivated in temperate gardens, mostly stemless perennials with fleshy corms. The leaves are distinctive, typically trifoliolate with fleshy petioles, and with a sour, astringent juice. The flowers are five-merous with five petals and ten or more stamens often held at different levels, and a superior ovary.

PITTOSPORUM (Pittosporaceae) A genus of some 210 species, a few species of *Pittosporum*, especially

Platanaceae, *Platanus ×acerifolia.* London plane is widely grown as a hardy street tree. The family is characterized by spirally arranged, palmately lobed leaves with the petioles swollen at the base and covering the axillary bud. The small, unisexual flowers are in rounded heads, the male flowers with a few minute sepals and a similar number of stamens, and the female flowers with a few sepals and several separate carpels. The achenes, each with a tuft of bristles, are massed in spherical heads.

those with variegated foliage, are cultivated in temperate gardens. The genus and family have simple leaves arranged spirally, five-merous flowers, five stamens, and a superior ovary with a single style. The Australian genus *Billardiera* comprises shrubs and vines that are half-hardy and have attractive drooping blue or white flowers, some with strikingly blue berries.

PLATANUS (Platanaceae) In a small family of just one genus of trees, *Platanus* is important in urban forestry. *Platanus* species are most often seen lining avenues and streets. *Platanus orientalis*, the Eurasian species, is an important source of building lumber as well as being ornamental. The hybrid between this species and the North American native *P. occidentalis*, called *P. ×acerifolia* (London plane), is widely used across the world in street plantings. Recognizable by its spirally arranged, palmately lobed leaves with soft indumentum, *Platanus* has small unisexual flowers, clustered in dense, spherical heads borne on drooping stalks. The flowers are wind pollinated, and the seeds are distributed by wind. Oddly enough, its closest living relatives are Nelumbonaceae (described following Nymphaeaceae in Families A–Z) and Proteaceae.

RESEDA (Resedaceae) A single annual species, *Reseda odorata* (mignonette), is occasionally grown for its delightful scent; the small flowers have an inconspicuous calyx and corolla but many prominent stamens. Other species of *Reseda* are annuals with attractive yellow or white flowers, as in *R. alba*, but are rarely cultivated. The genus and family are unusual flowering plants in having the single-celled ovary open at the apex and bearing minute stigmas on the rim.

RIBES (Grossulariaceae) A northern temperate genus of evergreen and deciduous shrubs, often spiny, *Ribes* includes gooseberries and currants, and a few ornamental shrubs. Flowers are small, usually drooping, and have a tubular hypanthium, small petals, and an inferior ovary. The fruits are berries, typically bearing the remains of the calyx and corolla at the tip. The genus was historically included in Saxifragaceae, but the structure of the flowers is quite different and molecular studies show them to be only distantly

Resedaceae, *Reseda alba*

related. Species include *R. nigrum* (black currant), *R. rubrum* (red currant), *R. sanguineum* (flowering currant), and *R. uva-crispa* (gooseberry).

SARRACENIA (Sarraceniaceae) One of three genera of carnivorous plants of this small family, *Sarracenia* species are distinguished by their pitcher-like leaves in which insects are trapped and digested. *Sarracenia* and *Darlingtonia*, a second genus, are occasionally grown for their unusual leaves, mottled and veined with yellow, purple, and red. Success in growing depends on using nonchlorinated water. The nodding flowers are quite large and borne on long stalks. Most species favor boggy habitats.

SHORTIA (Diapensiaceae) Hardy and cultivated in rock gardens, *Shortia* species are low-growing perennials of the northern temperate and Arctic family Diapensiaceae. The southeastern North American *Galax urceolata*, the only species of the genus and one of just 13 of the family, is sometimes grown as a ground cover. Characteristics of Diapensiaceae are five-merous flowers with petals partly united, five functional stamens, and five sterile stamens (staminodes). The superior, three-celled ovary consists of three united carpels with a single style and three-lobed stigma. Some *Shortia* species have distinctive, fringed petals.

STACHYURUS (Stachyuraceae) The only genus of the family, with five or six species, *Stachyurus* includes small deciduous trees and shrubs of eastern Asia that

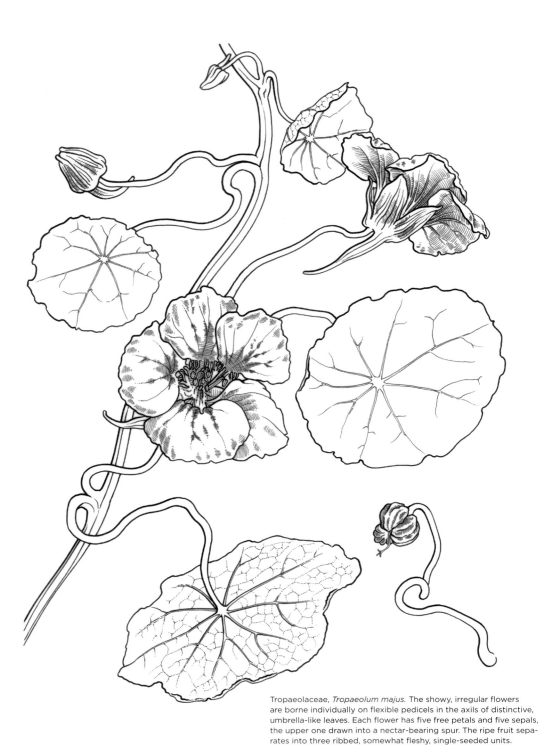

Tropaeolaceae, *Tropaeolum majus*. The showy, irregular flowers
are borne individually on flexible pedicels in the axils of distinctive,
umbrella-like leaves. Each flower has five free petals and five sepals,
the upper one drawn into a nectar-bearing spur. The ripe fruit separates into three ribbed, somewhat fleshy, single-seeded units.

flower in spring before the leaves appear. The flowers are borne in drooping spikes or racemes and are four-merous, with four overlapping petals and eight stamens. The superior ovary consists of four cells and a single style. Well-grown plants make a remarkable sight, blooming early in spring and covered with pendent inflorescences of small, pale to deep yellow or sometimes pink flowers. Most common in cultivation, *S. praecox* also has attractive fall foliage. Less frequently grown in gardens, *S. salicifolius* has narrow, willow-like leaves with the new foliage flushed pink or red. The common name for the genus, spiketail, alludes to the drooping inflorescences of closely clustered flowers.

TAMARIX (Tamaricaceae) A small family of five genera and about 80 species of Eurasia and Africa, Tamaricaceae comprise small trees and shrubs. Only the genus *Tamarix* is cultivated for its masses of small pink flowers and its willowy habit. The tiny leaves are best described as granular, recalling those of some *Cupressus* (cypress) species. The flowers, borne in spikes or panicles, are four- or five-merous and have a superior ovary. Some species are invasive in Australia and parts of North America.

TASMANNIA (Winteraceae) Belonging to the particularly archaic paleodicot family Winteraceae, the evergreen shrub or small tree *Tasmannia lanceolata* (widely known by its synonym, *Drimys lanceolata*) is occasionally grown as an ornamental, mainly for its attractive red stems. The male and female flowers are borne on separate plants and are inconspicuous and uninteresting display-wise. Winteraceae are largely Australasian and South American (one genus occurs in Madagascar) and have simple leaves and usually many stamens and free carpels. Fossils assigned to Winteraceae are reported from South Africa and California. *Tasmannia* occurs in Australia, New Guinea, and Indonesia. *Drimys* as understood today is exclusively South American.

TECOPHILAEA (Tecophilaeaceae) A half-hardy Chilean geophyte with attractive blue flowers, *Tecophilaea* is occasionally grown in rock gardens. A monocot with parallel-veined leaves, Tecophilaeaceae have three-merous flowers, six stamens, some often sterile, anthers with apical pores, and a half-inferior ovary; the capsules contain small black seeds. The genus *Cyanella* includes several very attractive bulbous species, unfortunately none in general cultivation.

TIBOUCHINA (Melastomataceae) With beautiful purple flowers, species of *Tibouchina* (glory bush) do not tolerate frost but thrive in warm temperate and Mediterranean gardens. The genus and family are almost entirely tropical and typically have simple, opposite leaves, usually with three or more main veins from the base, five-merous flowers, and stamens in two whorls of differently formed twisted filaments. The anthers often have terminal pores, and the anther connectives may have appendages. Species of a few other genera of Melastomataceae are occasionally cultivated, including *Dissotis*, which have attractive pink or purple flowers.

TROPAEOLUM (Tropaeolaceae) The garden nasturtium, *Tropaeolum majus*, a favorite across the world, is a member of the family Tropaeolaceae, with just one genus, of Central and South America. All are trailing plants or vines with attractive flowers, but none is truly cold hardy. The distinctive umbrella-like leaves contain watery sap with a sharp taste reminiscent of the mustard family, especially of watercress, the genus *Nasturtium* (Brassicaceae), hence the common name nasturtium. Pickled fruit of *T. majus* can be used a substitute for capers. Unlike Brassicaceae, to which they are related, Tropaeolaceae have five sepals and five petals. The superior ovary consists of three cells, each with a single ovule. When ripe, the fruit separates into three units, often somewhat fleshy and each containing a single seed. The garden nasturtium is a cultigen of hybrid origin, most likely derived from wild species native to Peru.

GLOSSARY

ACAULESCENT: without a stem

ACCRESCENT: of floral parts enlarging during development of the fruit

ACHENE: small, dry, one-seeded fruit, not splitting when ripe

ALTERNATE: when of leaves, see spiral

ANDROGYNOPHORE: stalk in some specialized flowers bearing both stamens and carpels above the point of corolla or perianth attachment

ANNUAL: plant living for just one or two seasons in a year

ANTHER: pollen-containing structure, consisting of two lobes each containing two pollen sacs, these splitting in a consistent manner, usually by longitudinal slits

APPRESSED: lying flat against

AQUATIC: plant growing in water

ARIL: fleshy appendage or outgrowth of a seed, often red; arillate is the adjective

AXIL: angle between leaf and stem; axillary is the adjective

BERRY: fleshy fruit containing many seeds and with no hard layer containing each individual seed (for example, a tomato, *Lycopersicum*)

BIFACIAL: of leaves with identical surfaces and oriented edgewise to the stem, that is, not with an upper surface and a lower surface

BIPINNATE: twice pinnate

BRACT: very reduced leaf, often scale-like, often associated with the flower stalk or inflorescence; bracteate is the adjective

BULB: underground storage organ consisting of swollen leaf bases on a condensed stem axis, dormant at certain times of the year (for example, an onion, *Allium*)

CALYX: outer whorl of a flower when different from the corolla, usually green, consisting of sepals; calyces is the plural

CAPITATE: abruptly enlarged at the tip

CAPSULE: dry fruit derived from a compound ovary (consisting of more than one carpel) and splitting open when ripe

CARPELS: female organs of a flower, each consisting of an ovary, style (when present), and stigma; carpels may be free, variable in number, or two or more united in a compound structure, then with the ovary consisting of a single cell or as many cells as carpels and with a single style or as many styles as carpels

CATKIN: inflorescence of tiny flowers with scaly bracts, usually pendulous

COMPOUND: of leaves, a leaf consisting of multiple leaflets sharing the same petiole

CORM: underground storage organ consisting of swollen stem tissue, dormant at certain times of the year (for example, *Gladiolus*)

COROLLA: whorl of a flower just interior to the calyx, consisting of petals, often brightly colored

CORYMB: flat-topped inflorescence with several stalked flowers arising from different levels on the stem, the outer (lower) flowers opening first; corymbose is the adjective

CULTIGEN: a cultivated plant species, not known in the wild

CULTIVAR: a genetically uniform selection of a plant species or hybrid, usually registered and named

CYME: inflorescence in which the terminal flower is the first to unfold, followed by those on lateral branchlets; cymose is the adjective

DECIDUOUS: falling off seasonally or at a certain stage of development

DECUSSATE: arrangement of opposite leaves in alternate pairs at 90 degrees from the next

DIGITATE: of a compound leaf with more than three leaflets arising from the top of the petiole

DISSECTED: of a leaf divided into many slender segments

DISTALLY: away from the point of attachment

DRUPE: fleshy fruit with a single seed, with the inner layer hard and woody (for example, a peach)

ELAIOSOME: fleshy, often white appendage attached to a seed

EPIPHYTE: plant growing on braches of other plants rather than on the ground; epiphytic is the adjective

EXTRORSE: of anthers splitting open via longitudinal slits facing away from the center of the flower, best determined in cross section of a flower bud

FLORET: one of many small flowers in a crowded inflorescence (as in daisies, Asteraceae, and grasses, Poaceae)

FOLLICLE: dry fruit derived from a single carpel and containing multiple seeds, splitting down one side when ripe

GEOPHYTE: perennial plant progagating from an underground organ; geophytic is the adjective

HEAD: inflorescence with flowers crowded together at the top of the stem, usually without individual stalks

HYPANTHIUM: floral cup or tube bearing calyx, corolla, and stamens on the rim

IMPARIPINNATE: see pinnate

INFERIOR: adjective to describe the position of the ovary relative to the rest of the flower (an important character for distinguishing many families), thus the perianth and stamens borne at the top of the ovary; compare with superior

INFLORESCENCE: a cluster of flowers on a single axis

INTRORSE: of anthers splitting open via longitudinal slits facing the center of the flower

LOCULE: a chamber of the ovary

-MEROUS: referring to the number of parts, as in flowers four-merous, parts in multiples of four

NUT: one-seeded, dry fruit not splitting when ripe

NUTLET: small nut-like fruit

OBSOLETE: reduced to a vestige or lacking

OPPOSITE: arrangement of leaves or floral parts in a pair on either side at the same level; compare with spiral and whorl

OVARY: portion of the carpel that contains the ovules

OVULE: egg-containing structure within the ovary

PACHYCAUL: growth habit of tree-like plants with a disproportionately thick stem and few if any branches

PALMATE: of the arrangement of veins with multiple main veins from the base of the leaf blade, or of leaf shape with lobes arranged in radiate or hand-like manner

PANICLE: an indeterminate inflorescence of many branches in apparently complex arrangement; paniculate is the adjective

PARIPINNATE: see pinnate

PEDICEL: stalk of an individual flower in an inflorescence

PEDUNCLE: stalk of an entire inflorescence

PENDENT: hanging

PERENNIAL: long-lived plant dying back completely in winter, resprouting in spring, usually not woody above the ground

PERFECT: of a flower with both female and male organs present

PERIANTH: the two nonreproductive outer whorls of a flower, often used when the calyx and corolla are not differentiated in color, texture, etc., from one another

PERSISTENT: remaining attached, not falling with age

PETALOID: petal-like

PETALS: units of the corolla, sometimes joined together, then sometimes called lobes

PETIOLE: stalk of a leaf

PINNATE: of a leaf with multiple leaflets sharing the same petiole, thus arranged like a feather, which is paripinnate, with an equal number of leaflets (without a terminal leaflet), or imparipinnate, with an unequal number of leaflets (thus with a terminal leaflet), or of a lobed leaf shaped in a similar fashion but not divided into separate leaflets

POLLEN: the grains borne within the anther that contain the male reproductive cells

RACEME: simple, elongate inflorescence with the flowers opening in sequence beginning from the bottom, and each flower with individual stalk, the pedicel

RADIALLY SYMMETRIC: of a flower that can be bisected in multiple vertical planes to yield matching halves

RECEPTACLE: enlarged end of a stem on which the flower parts are borne

RESUPINATE: twisted through 180 degrees, thus upside down (for example, flowers of lobelioid Campanulaceae and orchids)

RHIZOME: underground (or surface) creeping storage organ of stem origin, with obvious nodes, dormant at some time of the year (often winter), with terminal leaves when growing (for example, *Iris*)

SAMARA: winged, dry fruit not splitting when ripe

SEPALS: units of the calyx, sometimes joined together, then often called calyx lobes

SESSILE: of organs without a stalk

SHRUB: woody plant of modest size with multiple stalks from base (less specifically, any small tree)

SIMPLE: of leaves with a single blade attached to the petiole (or petiole sometimes absent)

SPADIX: thick or fleshy spike, typically subtended by a large spathe

SPATHE: bract, sometimes colored, that subtends and envelops a spadix or other types of inflorescence

SPIKE: inflorescence with sessile (stalkless) flowers opening in sequence beginning from the bottom

SPIKELET: secondary spike, in a compound inflorescence that is itself a spike

SPIRAL: arrangement of leaves or floral parts not opposite or in a whorl, thus placed singly at different levels; also called alternate, referring to leaves

SPUR: tubular or sac-like projection of a petal or sepal, usually containing nectar

STAMENS: male organs of a flower, each with stalk (filament) and pollen-bearing structure (anther)

STAMINAL COLUMN: a column of stamens united basally or for some distance above the perianth

STAMINODE: sterile stamen, sometimes resembling a petal

STEM: main axis of a plant, bearing leaves, buds in leaf axils, and flowers

STIGMA: part of the carpel at the apex of the style, the site of pollen deposition, often sticky

STIPULE: basal appendage of a leaf, usually in pairs, often attached to the petiole, leafy or scale-like, sometimes protecting a bud and falling as the leaf unfolds; stipules of roses and pelargoniums are green and attached to the base of the petiole; stipulate is the adjective

STYLE: slender part of the carpel and bearing the stigma, usually borne at the top of the ovary, sometimes single or more than one per flower

SUBSHRUB: a low shrub, thus woody and remaining above the ground even when dormant

SUPERIOR: adjective to describe the position of the ovary relative to the rest of the flower (an important character for distinguishing many families), thus the perianth and stamens arising below the ovary; compare with inferior

TEPALS: sterile units of the flower when calyx and corolla not differentiated from one another, then outer tepals and inner tepals instead of calyx and corolla

THYRSE: compact inflorescence, often in axillary position, derived from condensed cymes

TRIFOLIOLATE: of a compound leaf bearing three leaflets on a common petiole

UMBEL: inflorescence with several (to many) stalked flowers arising from the same point and carried at the same level

UNIFACIAL: of leaves with two identical surfaces

UNITED: of parts joined together, as in petals joined together to form a single cup or tube

WHORL: arrangement of leaves or floral organs in a closed ring or circle; compare with opposite and spiral

ZYGOMORPHIC: of a flower that can be bisected in only one vertical plane to yield matching halves, thus bilaterally symmetric, often with lower petals or the tepals forming a lip and the upper somewhat hood-like

REFERENCES

Angiosperm Phylogeny Group. 2003. An update of the Angiosperm Phylogeny Group classification for the orders and families of flowering plants: APG II. *Botanical Journal of the Linnean Society* 141: 399–436.

Angiosperm Phylogeny Group. 2009. An update of the Angiosperm Phylogeny Group classification for the orders and families of flowering plants: APG III. *Botanical Journal of the Linnean Society* 161: 105–121.

Bell, A. D. 2008. *Plant Form: An Illustrated Guide to Flowering Plant Morphology, New Edition.* Timber Press.

Coombs, A. J. 2012. *The A to Z of Plant Names.* Timber Press.

Lawrence, G. H. M. 1951. *Taxonomy of Vascular Plants.* Macmillan.

Mabberley, D. J. 2017. *Mabberley's Plant-Book, Fourth Edition.* Cambridge University Press.

Stearn, W. T. 1992. *Botanical Latin, Fourth Edition.* Timber Press.

INDEX

aubergine.
    See *Solanum melongena*, 259
Aubrieta, 90
Aubrieta deltoidea, 89
Aucuba, 123, 124
Aucuba japonica, 124
Aucuba japonica 'Variegata', 124
Aulax, 229
Aurinia, 90
Aurinia saxatilis, 87
Avicennia, 21
avocado.
    See *Persea* and *P. americana*, 172
Azadirachta indica, 279
azalea and *Azalea*.
    See *Rhododendron*, 130, 131
Azara, 250
Azara microphylla, 250

B

Babiana, 166
baby blue-eyes.
    See *Nemophila*, 86
baby's breath.
    See *Gypsophila*, 107
Bacopa, 257
Bactris, 59
Baeometra, 118
balloon flower.
    See *Platycodon grandiflora*, 95
Ballota, 171
balsam, friar's.
    See *Styrax*, 262
Balsaminaceae, 278
bamboo.
    See Poaceae subfamily
        Bambusoideae, 216, 217
bamboo, heavenly.
    See *Nandina domestica*, 78
banana.
    See *Musa*, 279
banana family.
    See Musaceae, 279
Banksia, 229
baobab.
    See *Adansonia*, 183

baobab family.
    See Malvaceae, 183
Baptisia, 137, 140
Barberetta, 150
barberry.
    See *Berberis*, 76
barberry family.
    See Berberidaceae, 76
Barklya, 195
Barleria, 22
barley.
    See *Hordeum*, 217
basil.
    See *Ocimum basilicum*, 171
Basilicum, 171
basket of gold.
    See *Aurinia saxatilis*, 87
bastard box.
    See *Polygala chamaebuxus*, 219
Bauhinia, 137
Bauhinia galpinii, 138
bay, bull.
    See *Magnolia grandiflora*, 177
bayberry.
    See *Morella californica*, 142
bay tree.
    See *Laurus nobilis*, 172
bean.
    See *Phaseolus*, 139
bean, broad.
    See *Vicia faba*, 139
bear's breeches.
    See *Acanthus*, 21
bear's claws.
    See *Acanthus*, 21
beauty bush.
    See *Callicarpa*, 170–171,
        and *Kolkwitzia*, 10
Beauverdia, 34
bee balm.
    See *Monarda*, 171
beech.
    See *Fagus*, 141
beech, southern.
    See *Nothofagus*, 142

beech family.
    See Fagaceae, 141
beet.
    See *Beta*, 35
beet, sugar.
    See *Beta vulgaris*, 35
Begonia, 73
begonia, cane-stemmed.
    See *Begonia undulata*, 73, 74
Begonia boliviensis, 73
Begoniaceae, 73
begonia family.
    See Begoniaceae, 73
Begonia grandis, 73
Begonia sutherlandii, 73, 75
Begonia undulata, 73, 74
Belamcanda chinensis.
    See *Iris domestica*, 166
Bellevalia, 157
bellflower family.
    See Campanulaceae, 95
Bellis, 72
bells of Ireland.
    See *Moluccella*, 170
Beloperone.
    See *Justicia*, 21, 22
Berberidaceae, 76
Berberis, 76, 78
Berberis darwinii, 76
bergamot.
    See *Citrus*, 249
Bergenia, 252
Bergenia cordifolia, 252
Berkheya glabrata, 71
Berula, 47
Beschorneria, 64
Beta, 35, 37
Betula, 79
Betulaceae, 79
Bielschmiedia, 172
Bignonia, 82
Bignoniaceae, 81
bignonia family.
    See Bignoniaceae, 81
Billardiera, 281
Billia, 156

bindweed.
    See *Convolvulus*, 121
bindweed, blue rock.
    See *Convolvulus sabatius*, 121
bindweed family.
    See Convolvulaceae, 121
birch.
    See *Betula*, 79
birch family.
    See Betulaceae, 79
bird of paradise.
    See *Strelitzia reginae*, 274, 275
bird of paradise family.
    See Strelitziaceae, 275,
        following Zingiberaceae
birthwort.
    See *Aristolochia clematitis*, 60
birthwort family.
    See Aristolochiaceae, 60
bistort.
    See *Persicaria bistorta*, 222
bitterroot.
    See *Lewisia*, 224
blackberry.
    See *Rubus*, 239
black boys.
    See *Kingia* and *Xanthorrhoea*, 68
blanket flower.
    See *Gaillardia*, 72
blazing star.
    See *Liatris*, 72
Blechnum, 211
bleeding heart.
    See *Lamprocapnos*
        spectabilis, 207, 209
Blepharis, 22
Bletilla, 203, 205
blood grass, Japanese.
    See *Imperata cylindrica*, 216
bloodroot.
    See *Sanguinaria*, 211
bloodroot family.
    See Haemodoraceae, 150
bluebell.
    See *Campanula rotundifolia*, 95

clover, elk.

    See *Aralia californica*, 57

Cobaea, 218

Cobaea scandens, 218

cockscomb.

    See *Celosia*, 35

cocoa.

    See *Theobroma*, 183

coconut.

    See *Cocos nucifera*, 59

Cocos, 59

Cocos nucifera, 59

Coffea, 243

Coffea arabica, 243

Coffea robusta, 243

coffee.

    See *Coffea arabica* and

        C. robusta, 243

coffee family.

    See Rubiaceae, 243

coffee tree, Kentucky.

    See *Gymnocladus dioicus*, 139

Cola, 183

Colchicaceae, 117.

    See also under Liliaceae, 173

Colchicum, 117, 118

colchicum family.

    See Colchicaceae, 117

Coleonema, 247, 249

coleus.

    See *Plectranthus*

        scutellariodes, 171

Colletia, 236

Collomia, 218

Colocasia, 55

Colubrina, 236

comfrey.

    See *Symphytum*, 83

Commelina, 119

Commelina africana, 119

Commelinaceae, 119

Compositae.

    See Asteraceae, 69

confetti bush.

    See *Coleonema*, 247

Conicosia elongata, 30

Conium, 47

Conium maculatum, 47

Conophytum, 29

Conostylis, 150

Convallaria, 61, 64

Convallariaceae, 173.

      See also under Liliaceae, 173

Convolvulaceae, 121

Convolvulus, 121, 122

Convolvulus arvensis, 121

Convolvulus cneorum, 121, 122

Convolvulus coelisyriaca, 122

Convolvulus mauritanicus.

    See *C. sabatius*, 121

Convolvulus sabatius, 121

Convolvulus soldanella, 121

Coprosma, 243

Coptis, 235

coral tree.

    See *Erythrina*, 137

coral vine.

    See *Antigonon leptopus*, 222

Corchorus, 183, 184

Cordia, 83

Coreopsis, 72

coriander.

    See *Coriandrum sativum*, 43

Coriandrum, 47

Coriandrum sativum, 43

corn.

    See *Zea mays*, 217

Cornaceae, 123.

    See also under *Corokia*, 277

cornel, dwarf.

    See *Cornus canadensis* and

        C. suecica, 123, and *C.*

        unalaschkensis, 123, 124

cornflower.

    See *Centaurea cyanus*, 149

cornflower, Australian.

    See *Brunonia australis*, 149

Cornus, 123, 124

Cornus canadensis, 123

Cornus controversa, 124

Cornus florida, 123

Cornus kousa, 123

Cornus mas, 123, 124

Cornus nuttallii, 123, 124

Cornus suecica, 123

Cornus unalaschkensis, 123, 124

Corokia, 277

Corokia cotoneaster, 277

Coronilla, 140

Correa, 249

Cortaderia, 216, 217

Cortusa, 226

Corydalis, 207, 211

Corydalis elata, 207

Corydalis flexuosa, 207

Corydalis solida, 207

Corydalis triternata, 209

Corylopsis, 155

Corylopsis pauciflora, 153

Corylopsis spicata, 153

Corylus, 79, 141

Corymbia ficifolia, 189

Corymbia ficifolia, 190–191

Corynostylis, 270

Corypha, 59

Cosmos, 72

Costus, 275

Cotinus, 41

Cotinus coccyrigia, 41

Cotoneaster, 242

cotton.

    See *Gossypium*, 183

Cotyledon, 125

Cotyledon orbicularis, 126

cowslip.

    See *Primula*, 226

Crambe, 90

cranberry.

    See *Vaccinium*, 133

crane flower.

    See *Strelitzia reginae*, 274, 275

cranesbill.

    See *Geranium*, 148

cranesbill family.

    See Geraniaceae, 147

Crassula, 125

Crassulaceae, 125

Crassula columnaris, 125

Crassula ovata, 125

Crassula perfoliata, 127

Crataegus, 242

Craterostigma, 257

creeper, Virginia.

    See *Parthenocissus*

        quinquefolia, 271

creeping jenny.

    See *Lysimachia*, 9, 226

crepe myrtle.

    See *Lagerstroemia indica*, 176

crepe myrtle family.

    See Lythraceae, 176

Crescentia, 82

Crinodendron, 277

Crinodendron hookerianum, 277

Crinodendron patagua, 277

Crinum, 40

Crocanthemum, 112

Crocanthemum canadense, 112

Crocus, 165, 166

Crocus sativus, 166

Crocus vernus, 165

Crossandra, 21, 22

Crotalaria, 140

Croton, 136

Cruciferae.

    See Brassicaceae, 87

Cryoptotaenia, 47

Cryptocarya, 172

Cryptostegia, 50

cucumber.

    See *Cucumis sativus*, 73

cucumber family.

    See Cucurbitaceae, 73,

        following Begoniaceae

cucumber tree.

    See *Magnolia acuminata*, 177

Cucumis melo, 73

Cucumis sativus, 73

Cucurbitaceae, 73, following

    Begoniaceae

Cucurbita moschata, 73

Cucurbita pepo, 73

cumin.

    See *Cuminum cyminum*, 43

Cuminum, 47

Cuminum cyminum, 43

Cunonia capensis, 278

Cunoniaceae, 278

cup and saucer vine.

    See *Cobaea scandens*, 218

Cuphea, 176

Cupressus, 283

Curcuma, 275

Curcuma longa, 273

currant.

    See *Ribes*, 281, and *Vitis*
        vinifera, 272.

    See also under Saxifragaceae, 252

currant, black.

    See *Ribes nigrum*, 281

currant, flowering.

    See *Ribes sanguineum*, 281

currant, red.

    See *Ribes rubrum*, 281

Cuscuta, 121

Cussonia, 58

Cyanea, 97

Cyanella, 283

Cyanotis, 119

Cyclamen, 9, 226

Cyclamen persicum, 227

Cyclocarya, 167

Cyclospermum, 47

Cydonia, 239, 242

Cymbalaria, 257

Cymbidium, 203, 205

Cynanchum, 50

Cynara, 72

Cynara cardunculus, 72

Cynodon, 217

Cynodon dactylon, 216

Cynoglossum, 83, 86

Cypella, 166

Cyperaceae, 128.

    See also under Poaceae, 217

Cyperus, 128, 129

Cyperus papyrus, 128

Cyphia, 95, 97

Cyphostemma, 272

cypress.

    See *Cupressus*, 283

cypress vine.

    See *Ipomoea quamoclit*, 122

Cypripedium, 203, 205

Cyrtanthus, 40

Cyrtanthus epiphyticus, 40

Cyrtanthus flanaganii, 40

Cysticapnos, 211

Cytisus, 137, 140

Cytisus scoparius, 140

D

Daboecia, 131, 133

Dactylocapnos, 211

Dactylocapnos scandens, 207

Dactylorhiza, 203, 205

daffodil.

    See *Narcissus pseudonarcissus*, 39

Dahlia, 72

dahnia.

    See *Coriandrum sativum*, 43

Dais, 267

Dais cotinifolia, 267

daisy, Barberton.

    See *Gerbera*, 69

daisy, Namaqualand.

    See *Dimorphotheca* and
        Osteospermum, 69

daisy, Transvaal.

    See *Gerbera*, 69

daisy family.

    See Asteraceae, 69

Dalea, 140

Dampiera, 149

Danae, 64

Daphne, 267

Daphne bholua, 267

daphne family.

    See Thymelaeaceae, 267

Daphne genkwa, 267

Daphne laureola, 267

Daphne mezereum, 267

Daphne odora, 267

Daphne sericea, 267

Darlingtonia, 281

Darmera, 252

Darwinia, 191

Dasypogonaceae, 68, following
    Asphodelaceae

Datura, 259. 261

Datura stramonium, 259

Daucus, 43, 47

Davidia, 123, 124

Davidia involucrata, 124

daylily.

    See *Hemerocallis*, 68

daylily family.

    See Hemerocallidaceae, 68,
        following Asphodelaceae

Deinanthe, 161, 163.

    See also under Saxifragaceae, 252

Deinanthe bifida, 161

Delonix, 140

Delosperma, 29, 30

Delphinium, 233, 235

Dendrobium, 205

Dendromecon, 207, 211

Deutzia, 161, 163

Deutzia gracilis, 163

Deverra, 47

Dianella, 68

Dianthus, 107, 110

Dianthus angulatus, 108

Dianthus orientalis, 108

Diapensiaceae, 281

Diascia, 257

Diascia vigilis, 256

Dicentra, 207, 211

Dicentra spectabilis.

    See *Lamprocapnos*
        spectabilis, 207, 209

Dichelostemma, 64

Dichondra, 122

Dichondra micrantha, 122

Dichondra repens, 122

Dichorisandra, 119

Dichroa, 163

Dichroa febrifuga, 161

Dicliptera, 22

Dictamnus, 247, 249

Dictamnus albus, 247, 249

Didymeles, 92

Dieffenbachia, 55

Dierama, 166

Dietes, 165, 166

Dietes bicolor, 165

Dietes grandiflora, 165

Digitalis, 255, 257

Dilatris, 150

Dilatris viscosa, 151

dill.

    See *Anethum graveolens*, 43

Dimorphotheca, 69, 72

Dionaea, 278

Dionysia, 226

Dioscorea, 122

Dioscoreaceae, 122

Diosma, 249

Diospyros, 262

Diospyros ebenum, 262

Diospyros kaki, 262

Diospyros virginiana, 262

Dipelta, 101, 105

Diphyllaea, 78

Diplacus, 257

Diplacus aurantiacus, 257

Diplarrhena, 166

Diplopanax, 124

Dipsacaceae.

    See Caprifoliaceae, 101

Dipsacus, 101, 105

Dipsacus sativus, 101

Dipteronia.

    See *Acer*, 24

Disa, 205

Disa longicornu, 204

Disa uniflora, 204

Dischidia, 50

Discocalyx, 226

Disporum, 117, 118.

    See also under *Prosartes*, 64

Disporum flavens, 117, 118

Dissotis, 283

dittany.

    See *Dictamnus*, 247

dock.

    See *Rumex*, 222

mandarin.

    See *Citrus reticulata*, 249

Mandevilla, 50

Mandragora, 261

mandrake.

    See *Mandragora*, 261

Mangifera, 41

Manglietia.

    See *Magnolia*, 177, 181

mango.

    See *Mangifera*, 41

mango family.

    See Anacardiaceae, 41

mangrove.

    See *Sonneratia*, 176

Manihot, 136

Manihot esculenta, 135

Mapania, 129

maple.

    See *Acer*, 11, 15, 24

maple, full moon.

    See *Acer japonicum*, 24

maple, Japanese.

    See *Acer palmatum*, 24, 25

maple, sugar.

    See *Acer saccharum*, 24

maple, vine.

    See *Acer circinnatum*, 24

maple family.

    See Aceraceae, 24

marigold, marsh.

    See *Caltha palustris*, 233, 235

Mariscus, 129

Markhamia obtusifolia, 82

maroela.

    See *Sclerocarya*, 41

marrow.

    See *Cucurbita pepo*, 73

Marrubium, 171

marula.

    See *Sclerocarya*, 41

marvel of Peru.

    See *Mirabilis jalapa*, 192, 193

Massonia, 157

masterwort.

    See *Astrantia*, 43

Mastixia, 124

Matthiola, 87, 90

Maytenus, 111

Meconopsis, 207

Meconopsis aculeata, 210

Medeola, 175

Medicago, 139, 140

Medicago sativa, 139

Melaleuca, 191

Melandrium.

    See *Silene*, 107

Melanthiaceae, 185.

    See also under Liliaceae, 173

Melanthium, 185

melanthium family.

    See Melanthiaceae, 185

Melastomataceae, 283

Melia, 279

Melia azedarach, 279

Meliaceae, 279

Melianthaceae, 148, following

    Geraniaceae

Melianthus, 148

Melianthus major, 148

Melicytus, 270

Melissa, 171

Melittis, 171

Melliodendron, 262

Melocactus, 94

melon.

    See *Cucumis melo*, 73

Mentha, 171

Menziesia.

    See *Rhododendron*, 131

Mercurialis, 136

Merendera.

    See *Colchicum*, 118

Merremia, 122

Mertensia, 83, 86

Merwilla, 157

Merwilla plumbea, 157

Merwilla plumbea, 159

Mesembryanthemum, 30

Mespilus, 242

Metapanax, 58

Metapanax davidii, 58

Metapanax delavayi, 58

Metrosideros, 191

Meum, 47

Michauxia campanuloides, 97

Michelia.

    See *Magnolia*, 177, 181

Michelia figo.

    See *Magnolia figo*, 177

Miersia, 34

mignonette.

    See *Reseda odorata*, 281

milk bush, yellow.

    See *Euphorbia mauritanica*,

    134, 135, 136

milkmaids.

    See *Burchardia*, 117

milkweed family.

    See Apocynaceae, 49

milkweeds, true.

    See Apocynaceae subfamily

    Asclepiadoideae, 49

milkwort family.

    See Polygalaceae, 219

Milula, 34

Mimetes cucullatus, 230

Mimosa, 140

mimosoids.

    See Fabaceae subfamily

    Mimosoideae, 137

Mimulus, 257

Mina lobata.

    See *Ipomoea lobata*, 122

mint.

    See *Mentha*, 171

mint family.

    See Lamiaceae, 169

Mirabilis, 193

Mirabilis jalapa, 192, 193

Miscanthus chinensis, 216, 217

Mitella, 252

Moluccella, 170, 171

Monanthes, 125

Monarda, 171

Monardella, 171

Monechma.

    See *Justicia*, 22

monkey flower.

    See *Mimulus*, 257

monkey flower, sticky.

    See *Diplacus aurantiacus*, 257

Monnina, 219

Monopsis, 97

Monotropa, 131

Monsonia, 147, 148

Monsonia section *Sarcocaulon*, 147

Monsonia speciosa, 148

Monstera, 55

Montia, 224

Montiaceae, 224

moonflower.

    See *Ipomoea aculeata*, 122

mophead.

    See *Hydrangea macrophylla*, 161

Moraceae, 187

Moraea, 166

Moraea tulbaghensis, 166

Morella, 142

Morella californica, 142

Morella cerifera, 142

Morinda, 243

morning glory.

    See *Convolvulus* and

    Ipomoea, 121

morning glory, purple.

    See *Ipomoea indica*, 120, 121

morning glory family.

    See Convolvulaceae, 121

Morus, 187

Morus alba, 187

moss, English.

    See *Sagina procumbens*, 110

moss, Spanish.

    See *Tillandsia usneoides*, 91

Muehlenbeckia, 222

Muehlenbeckia axillaris, 222

Muehlenbeckia complexa, 222

mulberry.

    See *Morus*, 185

mulberry, white.

    See *Morus alba*, 187

mulberry family.

    See Moraceae, 187

orange, mock.

    See *Philadelphus*, 160, and

        P. coronarius, 161

orange, Osage.

    See *Maclura pomifera*, 187

Orbea, 50

Orbea variegata, 48

orchid, bamboo.

    See *Sobralia macrantha*,

        202, 203

orchid, bee.

    See *Ophrys*, 205

orchid, slipper.

    See *Cypripedium*, 203

Orchidaceae, 203

orchid family.

    See Orchidaceae, 203

Orchis, 205

Orchis tridentata, 204

oregano.

    See *Origanum*, 171

Oreomunnea, 167

Origanum, 171

Orlaya, 47

Ornithogalum, 157

Ornithogalum narbonense, 159

Ornithogalum rupestre, 159

Ornithoglossum, 118

Orobanchaceae.

    See under Scrophulariaceae, 257

Orobanche, 257

Orontium, 55

Orontium aquaticum, 53

Orphium, 143, 145

Orphium frutescens, 143, 144

Oryza, 217

Oscularia, 30

osier.

    See *Cornus*, 123

Osmanthus, 197, 198

Osteospermum, 69, 72

Osteospermum calcicola, 69, 70

Ostrya, 79

Ostryopsis, 79

Othonna, 72

Ourisia, 257

Ovidia, 267

Oxalidaceae, 279

Oxalis, 279

Oxydendrum, 131, 133

Oxypetalum, 50

P

Pachylarnax.

    See *Magnolia*, 177, 181

Pachysandra, 92

Pachysandra procumbens, 92

Pachysandra terminalis, 92

Paeonia, 235

Paeoniaceae, 235, following

    Ranunculaceae

pajama bush.

    See *Lobostemon*

        fruticosus, 83, 84

Palisota, 119

Paliurus, 236

palm, Canary.

    See *Phoenix canariensis*, 59

palm, date.

    See *Phoenix dactylifera*, 59

palm, oil.

    See *Elaeis guineensis*, 59

palm, Senegal.

    See *Phoenix reclinata*, 59

palm, traveler's.

    See *Ravenala*, 275

palm, windmill.

    See *Trachycarpus*, 59

Palmae.

    See Arecaceae, 59

palm family.

    See Arecaceae, 59

Panax, 58

Panax ginseng, 58

Panax pseudoginseng, 58

Panax quinquefolius, 58

Pancratium, 39, 40

Pandorea, 82

Panicum, 217

pansy.

    See *Viola*, 270

Papaver, 207, 209, 210, 211

Papaveraceae, 207

Papaveraceae subfamily

    Fumarioideae, 208

Papaveraceae subfamily

    Papaveroideae, 206

Papaver nudicaule, 206, 207

Papaver orientale, 207

papaw.

    See *Asimina triloba*, 277

paper tree, Chinese.

    See *Edgeworthia*, 267

paper tree, Chinese pith.

    See *Tetrapanax papyrifera*, 57

paperwhites.

    See *Narcissus tazetta*, 39

Paphiopedilum, 205

papilionoids.

    See Fabaceae subfamily

        Faboideae, 137

papyrus.

    See *Cyperus papyrus*, 128

Parahebe.

    See *Veronica*, 257

Paraquilegia, 235

Parasyringa, 198

Paris, 185

Parkia, 140

Parnassia, 111.

    See also under Saxifragaceae, 252

Parrotia, 155

Parrotia persica, 153

parsley.

    See *Petroselinum crispum*, 43

parsley, water.

    See *Oenanthe*, 47

parsley family.

    See Apiaceae, 43

parsnip.

    See *Pastinaca*, 43

Parthenocissus, 272

Parthenocissus quinquefolia, 271

Parthenocissus tricuspidata, 271

pasqueflower.

    See *Anemone pulsatilla*, 233

Passerina, 267

Passiflora, 213

Passifloraceae, 213

Passiflora coccinea, 213

Passiflora edulis, 212, 213

Passiflora incarnata, 213

Passiflora ligularis, 213

Passiflora lutea, 213

Passiflora minata, 213

Passiflora mucronata, 213

Passiflora quadrangularis, 213

passion flower.

    See *Passiflora mucronata*, 213

passion fruit.

    See *Passiflora edulis*, 212, 213

Pastinaca, 43, 47

Patersonia, 166

Paulownia, 255, 257

Paulowniaceae.

    See Scrophulariaceae, 255

Paulownia tomentosa, 255

Pavetta, 243

Pavetta revoluta, 245

Pavonia, 184

pawpaw.

    See *Asimina triloba*, 277

pea.

    See *Pisum*, 139

pea, black-eyed.

    See *Vigna*, 139

pea family.

    See Fabaceae, 137

peach.

    See *Prunus*, 239

peanut.

    See *Arachis*, 139

peanut-butter tree.

    See *Clerodendrum*

        trichotomum, 170

pear.

    See *Pyrus*, 239

pear, pink wild.

    See *Dombeya burgessiae*, 182

pear, prickly.

    See *Opuntia*, 93

pearlwort.

    See *Sagina procumbens*, 110

pecan.

    See *Carya illinoinensis*, 167

Pelargonium, 147, 148

Pelargonium cucullatum, 148

Pelargonium ×*hortorum*, 146, 147

Peltoboykinia, 252

Pemphis, 176

Pennisetum, 217

pennywort, Indian, 47

Penstemon, 257

Pentanisia prunelloides, 245

Pentas, 243

peony.

    See *Paeonia*, 235

pepper.

    See *Capsicum annuum*,

      259, and *Piper*, 41

pepper, Szechuan.

    See *Zanthoxylum*, 249

pepperbush, sweet.

    See *Clethra alnifolia*, 114

peppercorn tree.

    See *Schinus molle*, 41

peppercorn tree, Brazilian.

    See *Schinus terebinthifolius*, 41

Pereskia, 94

Pereskiopsis, 94

Perilla, 171

periwinkle.

    See *Vinca major* and *V. minor*, 49

periwinkle, Madagascar.

    See *Catharanthus roseus*, 50

Perovskia, 171

Persea, 172

Persea americana, 172

Persicaria, 222

Persicaria amplexicaulis, 222, 223

Persicaria bistorta, 222

Persicaria virginianum, 222

persimmon, Chinese.

    See *Diospyros kaki*, 262

persimmon, common.

    See *Diospyros virginiana*, 262

persimmon, Japanese.

    See *Diospyros kaki*, 262

Persoonia, 229

Petasites, 72

Petrea, 269

Petrea volubilis, 268

Petroselinum, 47

Petroselinum crispum, 43

Petunia, 259, 261

Peucedanum, 47

Peucedanum galbanum.

    See *Notobubon galbanum*, 47

Phacelia, 83, 86

Phacelia fimbriata, 86

Phalaenopsis, 205

Phaseolus, 139, 140

Philadelphus, 160, 161, 163

Philadelphus coronarius, 161

Philodendron, 55

Phlebocarya, 150

Phlomis, 170, 171

Phlomis bracteosa, 170

Phlomis fruticosa, 170

Phlomis lychnitis, 170

Phlox, 218

Phlox drummondii, 218

phlox family.

    See Polemoniaceae, 218

Phlox longifolia, 218

Phlox paniculata, 218

Phlox subulata, 218

Phoenix, 59

Phoenix canariensis, 59

Phoenix dactylifera, 59

Phoenix reclinata, 59

Phormium, 68

Phrymaceae, 255, 257, with

    Scrophulariaceae

Phygelius, 257

Phygelius capensis, 254, 257

Phylica, 236

Phylica plumosa, 237

Phyllanthaceae, 136, following

    Euphorbiaceae

Phyllostachys, 217

Phylodoce, 133

Physalis, 259, 261

Physalis peruviana, 259

Physalis philadelphica, 259

Physocarpus, 239

Physostegia, 171

Phyteuma, 95, 97

Phytolacca americana, 193

Phytolaccaceae, 193, following

    Nyctaginaceae

Phytolacca dodecandra, 193

Pieris, 131, 133

pigeonberry.

    See *Phytolacca americana*, 193

pig's ear.

    See *Cotyledon orbicularis*, 126

pigweed.

    See *Amaranthus*, 35

Pimelea, 267

Pimenta, 191

Pimenta officinalis, 191

pimento.

    See *Pimenta officinalis*, 191.

    See also under

      Calycanthaceae, 181

pimpernel, common.

    See *Anagallis arvensis*, 226

pimpernel, scarlet.

    See *Anagallis arvensis*, 226

Pimpinella, 43, 47

Pimpinella anisum, 43

pincushion.

    See *Leucospermum*, 229

pincushion, blue.

    See *Brunonia australis*, 149

pincushion flower.

    See *Scabiosa*, 10

pineapple.

    See *Ananas comosus*, 91

pineapple family.

    See Bromeliaceae, 91

pineapple grass.

    See *Astelia*, 64

pink.

    See *Dianthus*, 107

pink, sea.

    See *Armeria*, 215

Pinus banksiana, 12

Piper, 41

Piperaceae, 41

Piriqueta, 213

Pisonia, 193

pistachio.

    See *Pistacia*, 41

Pistacia, 41

Pistia, 55

Pisum, 139, 140

pitanga.

    See *Eugenia uniflora*, 189

pith paper tree, Chinese.

    See *Tetrapanax papyrifera*, 57

Pittosporaceae, 279.

    See also under Buxaceae, 92

Pittosporum, 279

plane, London.

    See *Platanus* ×*acerifolia*,

      280, 281

Plantaginaceae, 255, 257, with

    Scrophulariaceae

Plantago, 255, 257

plantain.

    See *Plantago*, 255

Platanaceae, 281

Platanus, 281

Platanus ×*acerifolia*, 280, 281

Platanus occidentalis, 281

Platanus orientalis, 281

Platycarya, 167

Platycodon, 97

Platycodon grandiflora, 95

Plectranthus, 171

Plectranthus scutellariodes, 171

Pleione, 203, 205

Plerandra, 58

Plerandra elegantissima, 58

Plerandra veitchii, 58

Pleurospermum candollei, 45

plum.

    See *Prunus*, 239

Plumbaginaceae, 215

Plumbago, 215

Plumbago auriculata, 214, 215

Plumbago capensis.

    See *P. auriculata*, 214, 215

plumbago family.

    See Plumbaginaceae, 215

smoke tree.

    See *Cotinus coccygria*, 41

Smyrnium, 47

snakeroot, seneca.

    See *Polygala senega*, 219

snapdragon.

    See *Antirrhinum*, 255

snapdragon, Cape.

    See *Nemesia*, 257

snowball tree.

    See *Viburnum*

        macrocephalum, 105

snowdrops.

    See *Galanthus*, 39

snowflakes.

    See *Leucojum*, 39

soapberry family.

    See Sapindaceae, 156, following

        Hippocastanaceae

soapwort.

    See *Saponaria*, 107

Sobralia macrantha, 202, 203

Solanaceae, 259

Solanum, 259, 261

Solanum crispum, 260

Solanum laxum, 258, 261

Solanum lycopersicum, 259

Solanum melongena, 259

Solanum panduriforme, 261

Solanum tuberosum, 259

Soldanella, 226

Solenostemon, 171

Solomon's seal.

    See *Polygonatum*, 64

Sonneratia, 176

Sophora, 140

Sorbaria, 242

Sorbus, 242

sorrel.

    See *Rumex acetosa*, 222

soybean.

    See *Glycine max*, 139

Spanish flag.

    See *Ipomoea lobata*, 122

Sparaxis, 166

Sparrmannia, 184

Spartium, 137, 140

Spathodea, 82

speedwell.

    See *Veronica*, 255, 257

Spergularia, 110

Sphaeralcea, 184

spider plant.

    See *Tradescantia*, 119

spiderwort family.

    See Commelinaceae, 119

Spigelia marilandica, 145

spiketail.

    See *Stachyurus*, 283

spinach.

    See *Spinacia*, 35

spinach, New Zealand.

    See *Tetragonia tetragonioides*, 30

Spinacia, 35, 37

spindle tree.

    See *Euonymus*, 111

spindle tree family.

    See Celastraceae, 111

Spiraea, 239, 242

Spiraea thunbergii, 242

spiraeoids.

    See Rosaceae subfamily

        Spiraeoideae, 239

Spirodela, 55

Spondias, 41

Sprekelia, 40

spring beauty.

    See *Claytonia* and *Montia*, 224

spurge, Allegheny.

    See *Pachysandra procumbens*, 92

spurge, Japanese.

    See *Pachysandra terminalis*, 92

spurge family.

    See Euphorbiaceae, 135

squash.

    See *Cucurbita pepo*, 73

squash, summer.

    See *Cucurbita pepo*, 73

squill.

    See *Scilla*, 157

Stachys, 170, 171

Stachys albotomentosa, 170

Stachys byzantina, 170

Stachytarpheta, 269

Stachyuraceae, 281

Stachyurus, 281

Stachyurus praecox, 283

Stachyurus salicifolius, 283

Stackhousia, 111

Stackhousiaceae.

    See Celastraceae, 111

Stackhousia monogyna, 111

Stapelia, 50

statice and *Statice*.

    See *Limonium*, 215

Stellaria, 107, 110

Stellaria media, 107

Stenanthium, 185

Stenocarpus, 229

Stenocarpus sinuatus, 228, 229

Stenomesson, 40

Sterculia, 184

Sterculiaceae.

    See Malvaceae, 183

Sternbergia, 40

Stevia, 72

Stewartia, 265

Stewartia serrata, 265

stinkwood.

    See *Ocotea bullata*, 172

stock.

    See *Matthiola*, 87

stonecrop family.

    See Crassulaceae, 125

stone plants.

    See *Conophytum* and

        Lithops, 29, 30

storksbill.

    See *Pelargonium*, 148

strawberry.

    See *Potentilla*, 13

Strelitziaceae, 275, following

    Zingiberaceae

Strelitzia reginae, 274, 275

Streptocarpus, 82

Streptopus, 175

Strobilanthes, 21, 22

Strobilanthes dyerianus, 21

Strobilanthes pentstemonoides, 21

Strophanthus, 50

Strophanthus speciosus, 50

Strumaria, 40

Struthiola, 267

Strychnos, 145

Styloceras, 92

Stylophorum, 211

Styracaceae, 262

Styrax, 263

styrax family.

    See Styracaceae, 262

Styrax japonica, 262, 263

Styrax obasia, 262

Styrax officinalis, 262, 263

sultana.

    See *Vitis vinifera*, 272

sumac.

    See *Rhus*, 40

sumac family.

    See Anacardiaceae, 41

summersweet.

    See *Clethra alnifolia*, 114

sundew.

    See *Drosera*, 278

sunflower.

    See *Helianthus*, 72

sun-rose, false.

    See *Halimium*, 112

Sutera, 257

Swainsonia, 140

sweet flag.

    See *Acorus*, 277

sweet pea.

    See *Lathyrus*, 137

sweetshrub, Chinese.

    See *Calycanthus chinensis*, 181

sweetspire, holly-leaved.

    See *Itea ilicifolia*, 155

sweetspire, Virginia.

    See *Itea virginica*, 155

Swertia, 145

Swietenia, 279

Sycopsis, 155

Symphoricarpos, 105

Symphytum, 83, 86

312   INDEX

© Lendon J. Porter

Peter Goldblatt is a leading expert on the iris family, having spent his life studying its taxonomy and evolution. He is now retired from the Missouri Botanical Garden in St. Louis. Author of many scientific papers and books, he is coauthor with John Manning of *The Color Encyclopedia of Cape Bulbs* and *The Iris Family: Natural History and Classification*.

© Minoru Tomiyama

John C. Manning is senior specialist scientist at the South African National Biodiversity Institute. Although he has studied diverse plant families, his research more recently focuses on the iris and hyacinth families, collaborating with Peter Goldblatt, investigating the biology of the African genus *Lapeirousia* and the systematics of *Gladiolus*. John and Peter have coauthored several books, including *Gladiolus in Southern Africa* and wildflower guides to the southern African flora.